MW00713536

IMAGES
of America

INSTITUTE FOR
ADVANCED STUDY

This aerial shot of the Institute for Advanced Study (IAS), taken in 2005, shows additions to the campus since its main building, Fuld Hall (center with tower), was erected in 1939. (Photograph by David Kennedy.)

ON THE COVER: The Institute for Advanced Study's most widely recognized faculty member, Albert Einstein, is pictured with chalk in hand, standing alongside the institute's second director, Frank Aydelotte (center), and mathematicians Oswald Veblen (right) and James W. Alexander (left of spine). Aydelotte succeeded the founding director, Abraham Flexner, in 1939 and served until 1947. (Courtesy of the Shelby White and Leon Levy Archives Center, Institute for Advanced Study.)

IMAGES
of America

INSTITUTE FOR ADVANCED STUDY

Linda G. Arntzenius

ARCADIA
PUBLISHING

Copyright © 2011 by Linda G. Arntzenius
ISBN 978-0-7385-7409-7

Published by Arcadia Publishing
Charleston, South Carolina

Printed in the United States of America

Library of Congress Control Number: 2010932746

For all general information, please contact Arcadia Publishing:
Telephone 843-853-2070
Fax 843-853-0044
E-mail sales@arcadiapublishing.com
For customer service and orders:
Toll-Free 1-888-313-2665

Visit us on the Internet at www.arcadiapublishing.com

This book is dedicated in fond gratitude to Louise Morse
for her encouragement and wisdom.

CONTENTS

ACKNOWLEDGMENTS

Whether observed from the Institute Woods or from inside Wolfensohn Hall during a public lecture or concert, the international community that is the Institute for Advanced Study (IAS) is a fascinating part of the histories of Princeton and New Jersey. I am grateful to IAS director Peter Goddard for allowing access to the archival material that made this pictorial history possible.

The staffs of the Shelby White and Leon Levy Archives Center and the Historical Studies-Social Science Library have been particularly generous with their time and knowledge. I have benefited from the expertise of Christine Di Bella, Erica Mosner, and Marcia Tucker.

I am also grateful to the staffs of several other organizations, which include the following: Jeanette Cafaro and Eileen Morales at the Historical Society of Princeton; George Hawley and James Lewis at the Newark Public Library; Nancy Janow at the South Orange Public Library; and AnnaLee Pauls of the Department of Rare Books and Special Collections, Princeton University Library. I am indebted to the following past and present members of the institute staff: Kate Ablutz, Jim Barbour, Christine Ferrara, Linda Geraci, Roberta Gernhardt, Jennifer Hansen, Arlen Hastings, Dinah Kazakoff, Hank Pannell, Allen Rowe, James C. Stephens, Caroline Underwood, Georgia Whidden, and Mary Wisnovsky; to those photographers who shared their work: Adam Ashforth, Pryde Brown, David Graham, Randall Hagadorn, Andrea Kane, Cliff Moore, Ulli Steltzer, Linda Troeller, W. Brinton Whitall; and to those who shared their personal collections: Gary Alvin, Dennis Berlin, Gaby Borel, George Dyson, Lily Harish-Chandra, Louise Morse, Ingrid Selberg, and Marina von Neumann Whitman. Thank you also to the following for their reminiscences: Robert Landau, Andrew Lenard, and Shirley Satterfield; to Richard D. Smith for introducing me to Arcadia Publishing; and to Henry S. Horn for his expert knowledge of the Institute Woods. This book would not have been possible without the patience and support of my editor, Erin Rocha, and the staff at Arcadia Publishing, to whom I extend my heartfelt gratitude.

Unless otherwise noted, all images appear courtesy of the Shelby White and Leon Levy Archives Center, Institute for Advanced Study.

INTRODUCTION

Its name is synonymous with genius. It has been called a "Scholar's Paradise," a "Utopia," "the university of universities," and even described as the "penthouse on top of the Ivory Tower." Located just a mile from downtown Princeton, the Institute for Advanced Study's parklike setting defines it as a world apart, a fitting workplace for the intellectual giants who account for its lofty mystique.

Founded in 1930, the Institute for Advanced Study–Louis Bamberger and Mrs. Felix Fuld Foundation was a great experiment at the start of the Great Depression. Three stories are woven into its history. The first features two modest millionaires and their desire for an enduring legacy, the second a visionary educator with high ideals, and the third a moment of history that forced a group of European intellectuals to seek refuge in the United States.

L. Bamberger and Company, one of the nation's largest department stores, was sold by its owners, New Jersey businessman Louis Bamberger and his twice-widowed sister Caroline Bamberger Fuld, on the eve of the 1929 stock market crash. Guided by education expert Abraham Flexner, they created the Institute for Advanced Study (IAS) for the "pursuit of advanced learning and exploration in fields of pure science and high scholarship to the utmost degree." Scholars would be invited to work there without regard to "accidents of race, creed, or sex." Endowed with an initial gift of $5 million, it ultimately received the equivalent of over $200 million, as valued today, from Louis Bamberger and Caroline Fuld.

In addition to a permanent faculty of the best and brightest in their fields, the IAS has provided time away from teaching to talented scholars from around the world, totaling over 7,000 in 80 years. Although it has always eschewed measures of productivity, the numbers of award winners associated with the institute indicate a remarkable influence compared to its size—22 Nobel laureates, a majority of Fields Medalists, and numerous winners of the MacArthur Genius grants and Wolf Prize.

Because advancement of knowledge is its primary purpose, IAS professors have rarely sought popular acclaim, and it has been said that the institute is better known in Paris than in Princeton. Contrary to misconception, it is not a government think tank nor is it a part of Princeton University, although the rationale for its being in Princeton was the possibility of cooperation between the two institutions.

The original conception was for a community of scholars, not of buildings—its capital was not to be impaired by expenditures for site, structures, or equipment. For years it existed in rented space on the campus of Princeton University. To an outsider, it would, and did, appear as part of the university, a misperception that endures to this day. However, the relationship between the two is a symbiotic one, and many IAS professors teach at the university.

The institute supports the disinterested pursuit of learning that is free from private or commercial interests. It fosters the sort of fundamental innovative thinking that, Flexner believed, is the route to knowledge. In his provocatively titled essay "The Usefulness of Useless Knowledge," published in *Harper's Magazine* in October 1939, Flexner argued for curiosity-driven "theoretic" or *seemingly* useless research: "Institutions of learning should be devoted to the cultivation of curiosity, and the less they are deflected by considerations of immediacy of application, the more likely they are to contribute not only to human welfare but also to the equally important satisfaction of intellectual interest, which may indeed be said to have become the ruling passion of intellectual life in modern times." The theory of relativity, he pointed out, would not have been possible without the abstract mathematics of non-Euclidean geometry.

The hallmark of the institute is that the focus of research is determined only by the intense curiosity of the individual professor who has been chosen to join the faculty and that of the member scholars who visit for varying periods, ranging from a few months to a few years. Its independence is maintained by its initial endowment and financial contributions from individual donors, private foundations, and various arms of the federal government such as the National Science Foundation and the National Endowment for the Humanities.

Because it is a community of scholars rather than a typical university (it does not award degrees and its members are at the postdoctoral level), its history is centered on the individuals who have formed it, not only its founders but also the individual men (and it is primarily men with few exceptions until recent years) who have shaped it. Aspects of the institute today stem from its originators, the publicity-shy Bamberger siblings, the idealistic and controlling Flexner, the wood-chopping mathematician Oswald Veblen, and the efforts of Albert Einstein and others to rescue Jewish scholars from Nazi Germany. Each of its directors too has left a mark. There have been eight in all, over an 80 year period: Abraham Flexner (1930–1939), Frank Aydelotte (1939–1947), J. Robert Oppenheimer (1947–1966), Carl Kaysen (1966–1976), Harry Woolf (1976–1987), Marvin Goldberger (1987–1991), Phillip A. Griffiths (1991–2003), and Peter Goddard since January 2004.

Thus, you will find chapters on Louis Bamberger and Caroline Fuld, Flexner, Einstein, J. Robert Oppenheimer, and John von Neumann and his postwar electronic computer project, as well as one on the institute's first school, the School of Mathematics. After Hitler became chancellor of Germany in January 1933, a generation of promising Jewish mathematicians looked for refuge across the Atlantic, and the Institute for Advanced Study became a lifeline in the migration of European scholars to the United States.

A chapter on campus life illustrates aspects of the institute that go on beyond the purely academic, offering a glimpse of its day-to-day operations, its staff, and its celebrations.

No portrait of the Institute for Advanced Study would be complete without mention of the Institute Woods. IAS occupies an 800-acre campus, of which some 75 percent is preserved as open space that provides some 200 bird species a resting place during annual migrations. The woods are much beloved by Princeton residents.

Since its beginnings, the IAS has evolved along lines loosely set down by Flexner, who drew inspiration from the great research universities of Europe. He envisioned a community of "like-minded" scholars who would be able to work on their own, with little formal administration. Bamberger, on the other hand, thought of the faculty as employees—coworkers—who should be treated fairly but should still be subservient to the board of trustees and the director. This initial tension was never really faced by Flexner, and it would lead to problems for successive directors Frank Aydelotte and J. Robert Oppenheimer and come to a head with Carl Kaysen. Subsequently, however, things seem to settle on a more even keel.

This modest photo-essay's selection of images attempts to yield a glimpse of the Institute for Advanced Study's origins, purpose, outstanding faculty, far-reaching influence, and its position in local, national, and international history. While you will find images culled from the Institute for Advanced Study's 80 years, the emphasis here is on the first five decades, from its founding until it reached the pattern of organization of four schools that has remained stable since the mid-1970s. The institute has grown from one school to four: the School of Mathematics, the School of Historical Studies, the School of Natural Sciences, and the School of Social Science.

Flexner thought the institute should be "small and plastic." Since its founding, programs within the schools have changed and new subjects of study have been added. Today its scope includes theoretical computer science, East Asian studies, and, most recently, theoretical biology. The Institute for Advanced Study remains remarkably true to Flexner's original vision and his belief in curiosity-driven research. It has proven itself capable of change and renewal.

One

BEGINNINGS

THANKS TO L. BAMBERGER
AND COMPANY

As early as 1890, Louis Bamberger had his sights set on New Jersey for his first retail business venture. It took him two years to build up the capital and scout out a location. After careful study of Newark, walking the streets and assessing foot traffic, he found what he was looking for at 147 Market Street. Granted it was not Broad Street, the city's main thoroughfare, but it was not too far from it.

With the help of his brother-in-law Louis M. Frank and expert salesman Felix Fuld, Bamberger bought stock from a failed business and set about building what would eventually become "L. Bamberger & Company, Newark, New Jersey: One of America's Great Stores," as was posted on billboards from New York City to Philadelphia.

Bamberger, Frank, and Fuld, who called themselves "the firm," were innovative thinkers. Price tags replaced haggling, no-questions-asked returns inspired customer loyalty, advertising slogans spread the word, late-night openings increased sales, and a moving stairway (escalator), installed in 1901, was the first in New Jersey. The customer was always right and "coworkers," as Bamberger referred to his employees, enjoyed clubs, musical entertainments, paid vacations, and health benefits with company-matched contributions.

By 1912, Bamberger's occupied an entire city block. By the mid-1920s, "Bams" was a Newark institution. The store pioneered commercial radio broadcasting with WOR, the precursor to the CBS network. Coworkers could take advantage of a library and on-site university extension classes from Rutgers, the State University of New Jersey. Sales personnel studied merchandising and took an annual two-day expenses-paid trip to New York City to check out the competition. Such investments paid off. In 1928, the store ranked fourth in the nation. L. Bamberger and Company sold for $25 million to R. H. Macy and Company of New York in June 1929, just a few months before the stock market crash that launched the Great Depression.

Born in Baltimore, Louis Bamberger (1855–1944) was one of six children of German Jewish immigrants Elkan Bamberger and Theresa Hutzler, who ran a dry goods wholesale business. He left school at 14, worked in one of the largest stores in Baltimore, and then moved to New York City to act as a buyer for several out-of-town companies. Louis Bamberger worked hard and saved enough money to set up his own business. The growing city of Newark had commercial potential, and Bamberger would become its foremost citizen and benefactor. He felt a profound debt to the city for the success of L. Bamberger and Company. When the store was sold, he personally distributed $1 million in checks, ranging from $1,000 to $20,000, to 236 of his longest serving employees. Bamberger's original intention was to locate the Institute for Advanced Study on the Bamberger/Fuld estate in South and East Orange, but Abraham Flexner persuaded him that it should be situated close to a major university.

In 1883, Caroline Bamberger (1864–1944) moved to Philadelphia from her family's home in Baltimore and married Louis Meyer Frank, with whom she ran a successful dry-goods business before moving to Newark to help establish L. Bamberger and Company. Three years after Frank's death, she became Mrs. Felix Fuld. In so doing, her name came to embody the entire founding partnership of L. Bamberger and Company.

Felix Fuld was as full of energy as his friend and business partner Louis Bamberger was reserved. When Fuld married Bamberger's sister Caroline in 1913, the three were already inseparable. Fuld was a dedicated philanthropist. He gave anonymously to community-based and Jewish charities. Soon after his sudden death from pneumonia in January 1929, Caroline Fuld and Louis Bamberger decided to sell the business and devote themselves to philanthropy.

Louis Bamberger sat for this portrait in high turn-of-the-20th-century fashion in the studio of photographer A. C. Flett, which was located on the Atlantic City boardwalk. The photograph was made into a postcard, as it was the custom of the day. The two women are thought to be his sisters Caroline and C. Lavinia Bamberger.

The *Newark Evening News*, June 7, 1930, announced the Bamberger and Fuld gift of $5 million for the Institute for Advanced Study, "the first of its kind in the country." Note that the announcement states that the new institution will be located in Newark or vicinity.

Following Jewish tradition, the extended Bamberger family gathered for dinner on Friday evenings. In this photograph, Felix Fuld can be seen standing first from the left, with Louis Bamberger standing third from the left.

In early 1930, Louis Bamberger and Caroline Bamberger Fuld went west for the winter. In this undated photograph, taken at the Arizona Biltmore Hotel, Bamberger is shown with sisters C. Lavinia Bamberger (left) and Caroline Bamberger Fuld. Bamberger remained a bachelor throughout his 88 years and lived with his widowed sister Caroline until his death in 1944. Fuld died several months after her brother.

In the first quarter of the 20th century, the Fuld-Bamberger home at 602 Center Street, South Orange, sat on a 33-acre estate. It boasted a small working farm, extensive gardens, and a fine home, the rear of which is shown above. Caroline and Louis Bamberger lived here for over three decades. (Courtesy of South Orange Library.)

The Fuld Estate benefited from Caroline Bamberger Fuld's gardening talents. Fuld is celebrated for her gift of 2,000 Japanese flowering cherry trees to Branch Brook Park in Newark in 1927. The park, which was listed on the National Register of Historic Places in 1981, draws thousands of visitors to the area every spring. (Courtesy of South Orange Library.)

Felix Fuld (left) and Louis Bamberger were shrewd businessmen who complemented each other in temperament. Although he was a quiet man who liked music, especially opera, and collecting the signatures of signers of the Declaration of Independence, the diminutive Bamberger dominated the company. Described as "Newark's Merchant Prince," he was personally shy of reporters but enthusiastic about the new field of public relations, hiring Walter Moler as chief publicist in 1911 to promote the store as conscientious and service-oriented. Fuld has been described as "the firm's driving force." He built a loyal team of executives and stayed ahead of the curve. Bamberger and Fuld were committed to public service and made substantial charitable contributions in support of hospitals such as Newark's Beth Israel Hospital, the Jewish Theological Seminary, youth clubs, scholarships, and museums.

Bamberger's expanded "Great White Store" opened on October 16, 1912. The new eight-story building cost $2 million and covered an entire city block between Halsey Street on the east, Market Street on the south, Washington Street on the west, and Bank Street on the north. Bamberger's grew further in 1921 and 1929, adding more floors above ground and below street level. (Courtesy of Newark Public Library.)

This photograph, taken on a May morning in 1914, shows the Bamberger's fleet of some three dozen delivery trucks, two-thirds horse-drawn and one-third motorized, parked along Halsey Street. (Courtesy of Newark Public Library.)

"All roads lead direct to the free Bamberger's parking area, 107 Plane Street, near Central Avenue," according to this advertisement touting the company's location. Advertising played a large role in the store's success. When Bamberger's had a sale, the whole town turned out. (Courtesy of Newark Public Library.)

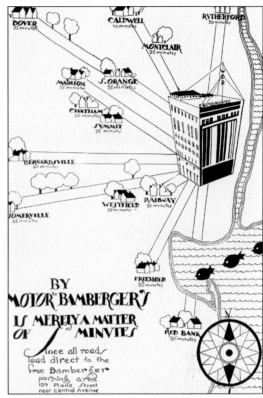

From horse-drawn carts to motorized vans, Bamberger's pioneered the home delivery of goods, as well as chauffeured customer service from its parking lot. In 1919, the store earned publicity for the first commercial use of an airplane, delivering an order of pajamas by parachute! (Courtesy of Newark Public Library.)

Louis Bamberger (center) broke ground for a new building for the Newark Museum, for which he had donated some $700,000, on Washington Street. Felix Fuld can be seen standing behind Bamberger, whose philanthropy to Newark was such that when he died, the flags of the city were flown at half-mast for three days and the city hall was decked in mourning. (Courtesy of New Jersey Historical Society.)

This 1940s aerial view of Newark shows the area occupied by L. Bamberger and Company. After it was sold to R. H. Macy and Company in 1929, Bamberger continued to serve as chairman of the board until 1939 and was a presence in the store until his death in March 1944. The store retained the Bamberger name into the 1980s. (Courtesy of Newark Public Library.)

Two

ABRAHAM FLEXNER
HIS DREAM

The commitment to learning and to philanthropy of L. Bamberger and Company's owners augured well for the man who would become the institute's first director, Abraham Flexner. Flexner had recently retired from an influential position with the Rockefeller Foundation when the opportunity to realize an ideal he had long held dear fell into his lap.

In late 1929, his reputation as *the* expert in the field of medical education brought him to the attention of Louis Bamberger and Caroline Bamberger Fuld shortly after their sale of L. Bamberger and Company.

The *Flexner Report* of 1910 was a highly influential critique of medical schools in the United States and Canada that had led to reforms in physician training. Hoping to found a medical school in Newark, the Bamberger siblings sought Flexner's advice. But if the wealthy philanthropists hoped Flexner would give life to their dream, they were mistaken.

Earlier in his career, Flexner had undertaken a similarly critical examination of American higher education. Over the years, he had developed his own dream of an advanced research institution in the United States, inspired by those in Europe. To Bamberger and Fuld, he described the need for an institution devoted to unrestricted research at the frontiers of knowledge. Persuaded by Flexner's reasoning, and with their own deep regard for learning, Bamberger and Fuld found their cause. They agreed to an initial endowment of $5 million with promises of more and insisted that Flexner become the new institution's first director. On May 20, 1930, a certificate of incorporation was filed with the State of New Jersey for the "Institute for Advanced Study—Louis Bamberger and Mrs. Felix Fuld Foundation."

Right from the start, the founders stipulated that the new institution would be free of prejudice as to race, religion, or sex. At a time when access to higher learning was restricted for women, for African Americans, and for Jews, this was an extraordinary statement. Serving as director until 1939, Flexner wrote, "I have sketched an educational Utopia. I have deliberately hitched the institute to a star."

Like Louis Bamberger, Abraham Flexner (1866–1959) came from a German Jewish immigrant family. His tradesman father, Moritz Flexner, and seamstress mother, Esther Abraham, had settled in Louisville, Kentucky, where Flexner was born the sixth of nine children. Though money was tight, the family scraped together enough to send Flexner and his older brother Simon to Johns Hopkins University. Both would hold influential positions: Simon as the first director of the Rockefeller Institute for Medical Research and Abraham as dispenser of Rockefeller largesse for medical and educational projects. In 1886, with a bachelor of arts in Classics from Johns Hopkins, Flexner returned to Kentucky and subsequently founded his own experimental school. The Institute for Advanced Study would be the culmination of his lifelong devotion to higher learning.

The first board of trustees of the Institute for Advanced Study comprised, from left to right, the following: (first row) Alanson B. Houghton, former U.S. ambassador to the United Kingdom; Caroline Fuld; Louis Bamberger; Florence R. Sabin of the Rockefeller Institute for Medical Research; and Abraham Flexner; (second row) Louis Bamberger's nephew Edgar S. Bamberger; attorney Herbert H. Maass; accountant Samuel D. Leidesdorf; Lewis H. Weed, dean of Johns Hopkins Medical School; attorney John R. Hardin; R. H. Macy and Company executive Percy S. Straus; Baltimore gastroenterologist Julius Friedenwald; Frank Aydelotte, president of Swarthmore College; and Alexis Carrel of the Rockefeller Institute. The one board member not pictured was Lt. Gov. (later governor) of New York Herbert H. Lehman. L. Bamberger served as board president, Fuld as vice president, Houghton as chairman, Leidesdorf as treasurer, and Aydelotte as secretary.

Born in Iowa, Oswald Veblen (1880–1960), the son of Andrew Anderson Veblen, a professor of mathematics and physics at the University of Iowa, pursued mathematics at Harvard University and the University of Chicago before joining the faculty of Princeton University in 1905. Like Flexner, Veblen regarded the lack of high-level scholarly research in the United States as a problem for the nation's future. Veblen's uncle, economist Thorstein Veblen, had written an influential book on the issue, *The Higher Learning in America,* and Veblen's own efforts earned him the title "statesman of mathematics." On reading news of the Institute for Advanced Study, Veblen wrote immediately to Flexner suggesting Princeton as the ideal location, of mutual benefit to both university and institute. Veblen was among Flexner's first appointments. (Courtesy of Louise Morse.)

Luther P. Eisenhart was chairman of the mathematics department at Princeton University when Flexner began gathering scholars for the new Institute for Advanced Study. With Eisenhart's approval, but much to the displeasure of others at the university, Flexner hired several of Princeton's top professors, offering higher salaries and better benefits to Veblen, James W. Alexander, and John von Neumann. (Courtesy of Princeton University Library.)

Princeton University's Fine Hall, which is known as Jones Hall today, housed the Institute for Advanced Study from 1933 until 1939, a fact that has led to the persistent confusion that the IAS is part of Princeton University. Brand new in 1931, Fine Hall had a handsome wood-paneled library and common room. It owed much of its successful design as a center for working mathematicians to Veblen. (Courtesy of Princeton University Library.)

Pictured in 1927 are Russian mathematician Paul A. Alexandroff (left) with Veblen and his wife, Elizabeth Richardson. Veblen loved the outdoors and promoted a wooded site in Princeton Township for the institute. The "wood-chopping professor" would lead groups of IAS members on brush-clearing expeditions in the woods. In 1957, the Veblens donated their Herrontown Woods property to Mercer County as a wildlife and plant sanctuary with walking trails. (Courtesy of Louise Morse.)

With the advent of the IAS, Princeton replaced Germany's Göttingen University as the new world center for mathematics, which in those days included physics. This 1930s photograph shows Veblen (left) with the German mathematician Edmund Landau (center) and Danish mathematician Harald Bohr. Bohr's physicist brother, the 1922 Nobel Laureate Niels Bohr, would pursue research at the IAS on several occasions between 1938 and 1958. (Courtesy of Louise Morse.)

Accountant Samuel D. Leidesdorf was a friend and business adviser to Louis Bamberger. Along with Herbert H. Maass, senior partner in the law firm that acted for R. H. Macy and Company, Leidesdorf approached Flexner on behalf of Bamberger and Mrs. Fuld to sound out the feasibility of their plans for a medical school in Newark. Leidesdorf would serve as IAS treasurer (1930–1957) and chairman of the board of trustees (1957–1968).

Frank Aydelotte (1880–1956), below right with Albert Einstein, was president of Swarthmore College when he was asked to join the Institute for Advanced Study's first board of trustees. In 1939, he succeeded Flexner as director and guided the institute through the war years until 1947. Born in Indiana, Aydelotte was a Quaker with a warm personality and a more inclusive style of leadership than his predecessor. (Courtesy of AFP, Getty Images.)

Albert Einstein took center place for this photograph that was taken on May 22, 1939, at the dedication ceremony for the institute's first building. Named for Caroline Bamberger Fuld and her late husband, Fuld Hall provided much-needed space. Since 1933, the IAS faculty had grown to include professors in economics, politics, and humanistic studies. While mathematicians had offices on the campus of Princeton University, most of the others had to make do with a remodeled

home on Alexander Street and various other rentals in the town. Pictured are, from left to right, Alanson B. Houghton, C. Lavinia Bamberger (Louis Bamberger's sister), Albert Einstein, Mrs. Abraham Flexner (the successful Broadway playwright Anne Crawford), Abraham Flexner, John R. Hardin, Herbert H. Maass, and Harold W. Dodds (president of Princeton University).

Designed by Jens Frederick Larson in the classical Georgian style he had used for Dartmouth College, Fuld Hall was the Institute for Advanced Study's first building and included faculty and administrative offices, a library, common room, lecture hall, and cafeteria. In the above photograph, the unmistakable figure of Einstein can be seen seated in the front row of the crowd gathered in celebration of the building's dedication in May 1939. The person giving the speech is presumed to be Abraham Flexner. In the photograph below, Einstein sits next to Herbert H. Maass. Flexner's brother Simon Flexner, director of the Rockefeller Institute for Medical Research, is the leaning figure that is seated next to the empty seat.

Abraham Flexner's step down from the doorway of the new Fuld Hall in May 1939 prefigures his stepping down as director of the Institute for Advanced Study later that same year. His leadership had been challenged by financial concerns, as well as by strained relations with the faculty and with the Bamberger siblings, who were noticeably absent from the dedication ceremony.

C. Lavinia Bamberger waits with trowel in gloved hand to symbolically set the Fuld Hall cornerstone. At the ceremony, Bamberger represented her siblings Louis Bamberger and Caroline Fuld.

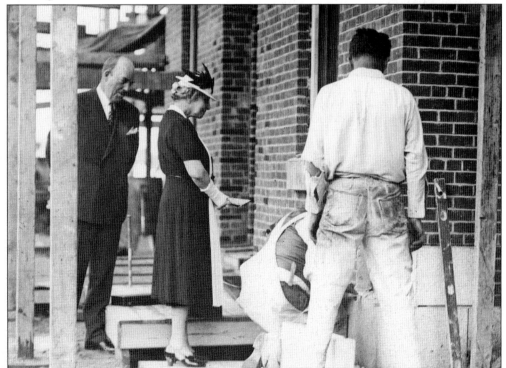

With the move to Fuld Hall, the IAS became "a world apart," as suggested by this 1940s photograph. Compared to the Princeton University campus, Fuld Hall was remote. Faculty and members gravitated back to Fine Hall for more than social contact. Institute mathematicians and physicists continued a symbiotic relationship with the university, holding most of their lectures and seminars there. These were open to advanced university students, just as university courses were open to IAS members. A similar relationship existed between the humanists and the university's Department of Oriental Studies and Department of Art and Archaeology.

Three

ALBERT EINSTEIN
ICH BIN FEUER UND FLAMME DAFÜR

Albert Einstein was in California when Abraham Flexner described his ideas for the Institute for Advanced Study. Aware of the worsening situation for Jewish scholars in his native Germany, Einstein was finding respite at the California Institute of Technology in Pasadena.

Since 1919, when his General Theory of Relativity had been confirmed by experiment, Einstein had been the world's most famous scientist. His findings had opened the door to an explosion of new physics including quantum mechanics.

But with Hitler's rising influence in Germany, the situation there for Einstein was becoming life-threatening. Einstein responded with enthusiasm to Flexner's repeated overtures. Of the IAS, he is reported to have said, "Ich bin feuer und flamme dafür," translated as "I am full of fire and flame for it."

On October 11, 1932, just eight months after Flexner's visit to California, the *New York Times* broke the news that Einstein had accepted an appointment at the "Scholar's Paradise" that was the Institute for Advanced Study. He remained at the IAS until his death in 1955.

For the next two decades, Einstein worked doggedly in search of a theory that would reconcile quantum mechanics and general relativity by unifying the four fundamental forces of nature, gravity, electromagnetism, the strong nuclear force, and the weak nuclear force.

In his leisure time, he played his violin with neighbors such as pianist Louise Strunsky and took his little second-hand sailboat out on Princeton's Lake Carnegie. He also used his considerable fame to publicize the plight of Jews in Germany and to promote pacifism and disarmament.

While Einstein's high profile was a constant source of concern for the controlling Flexner and the publicity-shy Bamberger and Fuld, his clear-eyed perspective on subjects beyond physics endeared him to the residents of Princeton. To this day, Einstein is fondly remembered by neighborhood children, students he invited into his home, shopkeepers, and members of the town's predominantly African American Jackson/Witherspoon Street neighborhood.

Albert Einstein (1879–1955) was visiting friends in Santa Barbara when this playful photograph was taken on February 6, 1933, shortly after he had accepted an appointment at the Institute for Advanced Study. In Berlin, Einstein's work was being denounced as "Jewish science," and he had offers from universities such as Leiden and Oxford. Attracted by Flexner's plans, Einstein came instead to the IAS. He arrived in Princeton to floods of demands from the press and charitable causes. Flexner thought he could "manage" his high-profile professor, but Einstein had his own interests and was not to be controlled. When Flexner refused an invitation from Pres. Franklin D. Roosevelt, on Einstein's behalf, the 1921 Nobel Laureate was furious. Einstein went to the White House in January 1934. (Courtesy of Archives; California Institute of Technology.)

Einstein's second wife, Elsa (center), with her daughter Margot (left) and Einstein's secretary Helen Dukas are pictured in this undated photograph in the IAS archives. Dukas accompanied the Einsteins to Princeton in October 1933 and lived with them in a rented duplex on Library Place for two years, until they moved to 112 Mercer Street. Einstein's stepdaughter Margot, who always called him "Uncle Albert," joined them in 1934. Albert and Elsa Einstein were first cousins.

Shortly after their arrival in Princeton, Elsa and Albert Einstein explored the countryside along the banks of the Delaware River with their friend Leon Watters (center) in 1934. Watters was a wealthy biochemist who ran a vocational school for Jewish boys in New York City. (Courtesy of Archives; California Institute of Technology.)

In 1947 in Princeton, Trude Fleischmann photographed Einstein's secretary since 1928, Helen Dukas (1896–1982), and the wife of historian Erich Kahler, Alice ("Lili") Kahler (right). The Kahler circle of friends included Einstein, Erwin Panofsky, Hetty Goldman, Ernst Kantorowicz, Thomas Mann, and painter Ben Shahn. After Elsa Einstein's death in 1936, Dukas, who never married, managed the Einstein household. When Einstein died, Dukas became his literary executor.

The house at 112 Mercer Street today looks much as it did in Einstein's day. The house, which passed to the IAS in 1986 after the death of Margot Einstein, remains a private residence owned by an IAS professor. Both professors who lived there since Einstein have been Nobel Laureates. In 2004, the institute donated items from the home to the Historical Society of Princeton. (Photograph by Dinah Kazakoff.)

In 1939, Maria ("Maja") Einstein (1881–1951) joined her brother in Princeton after anti-Semitic laws had been introduced in Italy, which was where she had been living with her husband, Paul Winteler. Because of ill health, Winteler was refused entry into the United States. The Einstein siblings had been close friends since childhood and over the years had grown to resemble one another a great deal. In 1946, Maja suffered a stroke from which she never recovered, and during his sister's long illness, Einstein was a loving companion, reading aloud to her every evening. Their friend Alice ("Lili") Kahler, who spent time with them in their rented summerhouse at Lake Saranac, recalls Einstein reading Herodotus. He refused to skip over sections that Kahler thought boring, just in case something important was missed!

The Indian statesman, Jawaharlal Nehru (right), visited Einstein at 112 Mercer Street in November 1949. Together with the spiritual and political leader Mohandas K. Gandhi, Nehru was a key figure in the Indian independence movement and was elected India's first prime minister in 1947. It is presumed that the two men discussed the fledgling state of Israel. At the urging of members of the Zionist movement, Einstein had written to Nehru two years earlier to ask for India's support for the Jewish homeland.

FULD HALL

In Princeton, Einstein spoke out for peace and justice. In the 1930s, he responded to countless requests from European Jews seeking refuge in America. In 1937, when African American contralto Marian Anderson was excluded from Princeton's whites-only Nassau Inn, Einstein invited her into his home. By the 1950s, he had acquired mythic status. Barbara Rose, his neighbor at 144 Mercer Street, was in grade school when he regularly walked by. Rose, whose favorite playground was the nearby stream in what was then known as "frog hollow," complained to her mother that "Dr. Iodine," as she called him, was allowed to wear sandals, while she had to wear boots. Her mother was unmoved. (Courtesy of Newark Public Library.)

Even in his 70s, regardless of weather, Einstein walked at least one way each day to his office in Fuld Hall. On occasion, he rode the IAS bus on his homeward journey. He never learned to drive and never owned an automobile. (Photograph by Alan W. Richards; courtesy of Fantova Collection of Albert Einstein; Department of Rare Books and Special Collections; Princeton University Library.)

In 1948, Einstein met with French-Columbian educator and politician Mario Laserna Pinzón to discuss a new university in South America. Those pictured are, from left to right, historian Dana C. Munro; scientist John von Neumann; classicist Whitney Oates; Einstein; game theorist Oskar Morgenstern; Laserna Pinzón; and mathematicians Samuel Wilks, Marston Morse, and Solomon Lefschetz. Laserna Pinzón helped create the Los Andes University in Bogota. (Courtesy of Louise Morse.)

This photograph of Einstein and the Hungarian physicist Leo Szilard (right) was taken at Lake Saranac just a month after the United States bombed Hiroshima in August 1945. Six years earlier, Szilard had persuaded Einstein to add his signature to a letter to Pres. Franklin D. Roosevelt warning of the dangers of Nazi Germany developing such a weapon. The letter led to the Manhattan Project and the first atomic bomb.

Einstein deplored the bombings in Japan and regretted signing the letter that had instigated the Manhattan Project. After the war, he advocated nuclear control. Einstein's neighbor Robert Strunsky worked for CBS and found Einstein receptive to the idea of broadcasting a speech from his home on the postwar dangers of atomic weapons.

Albert Einstein and Israeli Prime Minister David Ben Gurion (right) were photographed in the backyard of Einstein's home at 112 Mercer Street in 1952. Ben Gurion offered Einstein the presidency of the newly founded state of Israel. Einstein declined.

After Einstein received his first navy-colored cotton sweatshirt from Alice ("Lili") Kahler, it became his favorite garment. Kahler described Einstein's pleasure at receiving her gift and, at the same time, explained his abhorrence of socks as his "idiosyncrasy against wool."

Einstein is fondly remembered in Princeton for his unruly white hair, his delight in children, and his violin performances. He loved Mozart and Haydn but had no taste for Wagner. He loved to sail though he never learned to swim and never wore a life jacket. This photograph is featured in an informal Einstein museum in Landau, a family-run clothing store at 102 Nassau Street. (Courtesy of Gillett Griffin.)

In 1954, Agnes Lenard was a 26-year-old camp counselor at a Quaker-sponsored summer camp for refugees from the Ukraine in Lakewood, New Jersey, when she was invited to Einstein's home. Lenard is the young woman wearing glasses standing behind Einstein. In the mid-1960s and early 1970s, Lenard's brother Andrew, a mathematical physicist, was an IAS member. (Courtesy of Andrew Lenard.)

This photograph of Einstein's office at the Institute for Advanced Study, taken shortly after his death of heart failure in the Princeton hospital on April 18, 1955, captures the space as Einstein left it. Einstein's body was cremated, except for his brain, which was removed and preserved by the pathologist, Thomas S. Harvey, during the autopsy. Harvey was subsequently dismissed from his hospital position for refusing to give up Einstein's brain. Einstein's ashes were scattered at sea by his friend and executor Otto Nathan. Decades after his death, the world continues to be fascinated by the Institute for Advanced Study's most famous professor.

Four

SCHOLAR'S PARADISE
MATHEMATICS AND MORE

If, instead of a medical school, Bamberger and Fuld had originally planned to found a center for research in mathematics, their advisors may well have steered them towards Oswald Veblen rather than Abraham Flexner. The decision that Princeton was the best choice in New Jersey for the new institute was not only due to its first class library, but also in part because of Veblen's activities at Princeton University, where he had been realizing his own dream of a center for mathematical research for some years before the Institute for Advanced Study came to town.

Veblen guided the selection of professors for the institute's first school, as well as its policy of admitting only postdoctoral scholars. With the first professors installed in Fine Hall on the campus of Princeton University, Flexner witnessed his "educational Utopia" take shape in the form of Einstein and visiting Austrian physicist Erwin Schrödinger, Nobel Laureates both, in front of formulae-covered chalkboard. He reported finding some 60 university and IAS professors, students, and members gathered for afternoon tea, all "as happy as birds."

With his goal in mathematics achieved and in a hurry to move into other subjects, Flexner hired scholars in economics, art history, and archaeology. From the start, he had planned to create a series of autonomous schools or groups, but the man who had said the IAS would grow slowly, was running out of time.

By the mid-1930s, with Flexner almost 70 and Bamberger 80, the IAS had the following three schools: mathematics, economics and politics, and humanistic studies. Such rapid expansion highlighted the need for the IAS to have a place of its own, but this was the Great Depression, and money was tight even for the institute's wealthy patrons. Flexner struggled to meet the multiple demands of a growing number of professors and to appease his uneasy funders, who were tightening the purse strings.

Faculty & Workers

Albert Einstein. 1933.-

Oswald Veblen 1932-   Trustee. 1934-

James W. Alexander 1933-

Hermann Weyl 1933-

John v. Neumann 1933-

Walther Mayer 1933-

Charles C. Torrance 1933-1934, 1938-1939

John L. Vanderslice 1932-1935

Leo Zippin 1933-1936

Børge Jessen 1933-1934

Arnold N. Lowan 1933-1934

Thurman S Peterson 1933-1934

Meyer Salkova 1933-1934

P. R. von Kuryer 1933-1934

R. L. Wilder 1933-1934

A. Adrian Albert 1933-1934, Nov.1,1938

Kurt Gödel 1933-1934, 1935, 1938, 1940

D. H. Lehmer 1933-1934

Leonard M. Blumenthal 1933-1934, asst. 1935-1936

Harold S. Ruse 1933-1934

Willard E. Bleick 1933-1934, 1935

Tracy Yerkes Thomas 1933-1934

This opening page of signatures from the Institute for Advanced Study's first visitors' book includes several professors in the School of Mathematics, including Albert Einstein. While Einstein was a physicist, his field fell under the umbrella of mathematics at the time of the IAS founding. In 1966, a separate School of Natural Sciences was formed. Institute terminology for "faculty and workers" deserves comment. Professors such as Einstein are "faculty." Individuals supported for periods varying from a semester to one, two, or three years were "workers," called today "members." The now defunct category of "permanent member" was introduced by the third director, J. Robert Oppenheimer, for scholars with a continuing affiliation but no faculty appointment. Besides faculty, permanent members, and members, the IAS has had evolving categories over the years of visitor, director's visitor (also introduced by J. Robert Oppenheimer), and more recently, artist-in-residence.

In this undated photograph, IAS professor James W. Alexander (1888–1971) drives his Princeton University colleague Solomon Lefschetz, who lost both hands in an industrial accident in 1907. A Veblen protégé, Alexander was an accomplished mountaineer who is said to have left a window of his office on the top floor of Fine Hall ajar so that he could gain entry by climbing the building's exterior.

In addition to his pioneering work in topology, Veblen (right, with unidentified scholar) made significant contributions to the foundations of geometry and computing. "Do Not Erase" notices such as the one tacked to the chalkboard are still a common sight in the institute's hallways and common rooms, where similar boards offer an opportunity to record an important thought or formula. (Courtesy of Louise Morse.)

Hermann Weyl (1885–1955), pictured at left in 1950 and below (right) with David Hilbert in the late 1920s, left his native Germany to join the Institute for Advanced Study in 1933, three years after he had succeeded Hilbert to the world's most prestigious chair in mathematics at the University of Göttingen. Although Flexner had offered him an appointment in 1932, Weyl and his Jewish wife were reluctant to leave their homeland. When Hitler became chancellor in 1933, however, they recognized the deteriorating situation. Regarded as one of the most important mathematicians of the 20th century, Weyl is often quoted for his remark "My work always tried to unite the true with the beautiful, but when I had to choose one or the other, I usually chose the beautiful." (Left, courtesy of Annmarie Weyl Carr.)

Hungarian-born John von Neumann (1903–1957) was the institute's youngest professor in 1933 and was often mistaken for a graduate student. John von Neumann's contributions to science range across pure and applied mathematics, quantum mechanics, and economics. He pioneered game theory and convinced the IAS to build a prototype computer in the 1940s. As fun loving as he was brilliant, he was known as "good-time Johnny" for his legendary cocktail parties. This 1930s New Year's Eve party, above, shows, from left to right, the following: Angela Robertson, wife of Princeton University physicist Howard P. ("Bob") Robertson; von Neumann's first wife Mariette Kovesi; Eugene Wigner and his first wife, Amelia Frank; John von Neumann; an unidentified man; and Robertson. (Courtesy of Marina von Neumann Whitman.)

Austrian physicist Wolfgang Pauli (1900–1958) found refuge during World War II at the Institute for Advanced Study, where he was an early member and where he spent the years of World War II. In 1945, Pauli won the Nobel Prize in physics for his Pauli Exclusion Principle, the law that no two electrons in an atom can occupy the same quantum state simultaneously.

Marston Morse (1892–1977), who became a professor in 1935, is pictured (right) in Vienna in 1936 with Karl Menger, who started a mathematical colloquium at the University of Vienna in 1928. The colloquium was similar to the famed Vienna Circle, and IAS professors, including Kurt Gödel, John von Neumann, and Marston Morse, were frequent participants. (Courtesy of Louise Morse.)

This 1940s photograph shows, from left to right, James W. Alexander, Marston Morse, Albert Einstein, Frank Aydelotte, Hermann Weyl, and Oswald Veblen. Aydelotte directed the IAS from 1939 until 1947, during the years of World War II, a period when many of the faculty drew upon their own particular expertise to serve as consultants to the U.S. military.

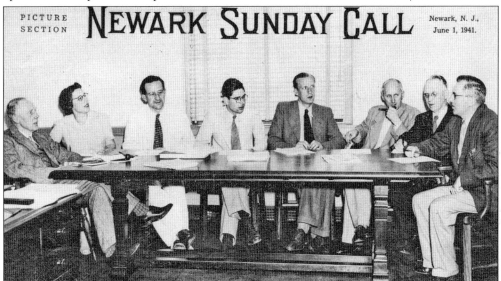

As recorded in this June 1941 issue of the *Newark Sunday Call*, Frank Aydelotte helped to bring a small, international group of economists and statisticians to Princeton with their families. At a time when the future of the League of Nations was unsure, IAS professors shared their offices, and secretaries worked in corridors in order to accommodate members of its economics section in Fuld Hall.

Frank Aydelotte's style as director was very different from that of his predecessor. He called on the faculty for advice and was generally respected and admired. According to Marston Morse, "Those of us who knew him best respected him and loved him." This 1940s photograph shows, from left to right, Alexander, Einstein, Aydelotte, Veblen, and Morse, who is fondly remembered for his support of generations of mathematicians. He has been quoted for the following remark: "Out of an infinity of designs a mathematician chooses one pattern for beauty's sake and pulls it down to earth."

Kurt Gödel (1906–1978) is arguably the most significant logician since Aristotle. His "incompleteness theorem" is a landmark of 20th-century formal logic. Among the first members in 1933, he was appointed professor in 1953. With physicist Julian Schwinger, Gödel received the IAS Einstein Award in 1951, which comprised a gold medal and $15,000, provided by trustee Adm. Lewis L. Strauss and his wife, Alice Hanauer Strauss. In the above photograph are, from left to right, Einstein, Strauss, Gödel, and Schwinger. Einstein and Gödel were photographed (below) by the economist Oskar Morgenstern. Morgenstern recalled that Einstein had once jokingly stated that his purpose for going to Fuld Hall was so that he could "have the privilege of walking home with Gödel." In later life, Gödel suffered an obsessive fear of food poisoning and, ultimately, died of malnutrition in Princeton Hospital. He is buried in the Princeton cemetery. (Above, photograph by Alan W. Richards; below, courtesy of Dorothy Morgenstern Thomas.)

Charles Rufus Morey of the Department of Art and Archaeology at Princeton University was the man to whom Flexner turned for advice on setting up the School of Humanistic Studies, later called the School of Historical Studies. The first appointment was classical scholar Benjamin D. Meritt, a specialist in Greek history, epigraphy, and archaeology; the second was renowned German art historian Erwin Panofsky. (Courtesy of Princeton University Library.)

Benjamin D. Meritt (1899–1989) was appointed professor in Greek epigraphy in 1935. That same year, Meritt (right), with German classicist Johannes Kirchner (left) and unidentified colleague, conducted field work on Greek's Attic peninsula. Meritt's study of the IAS collection of squeezes (copies of stone inscriptions made using pressed paper) as well as his over 200 scholarly articles and 14 books illuminate life in ancient Athens. (Courtesy of Christian Habicht.)

Erwin Panofsky (1892–1968) was appointed professor in 1935. Two years earlier, on a visit to New York City, Panofsky had received a telegram dismissing him from his post as head of the art history department at the University of Hamburg. The Nazi regime was excluding Jews from government office. Panofsky is considered one of the most influential art historians of the 20th century for his professionalization of the field in the United States. His extensive publications include *Studies in Iconology* (1939), *The Life and Art of Albrecht Dürer* (1943), and *Pandora's Box: the Changing Aspects of a Mythical Symbol* (1956), which was written with his first wife, Dora Mosse. The above photograph, taken in July 1930, shows Panofsky, who is seated cross-legged, with his Hamburg students during the annual two-week study trip that was part of the German system Flexner so admired. Dora Mosse is pictured, wearing a white shirt, behind her husband. (Courtesy of Irving Lavin.)

In 1936, archaeologist Hetty Goldman (1881–1972) became the institute's first female professor at a time when most universities did not admit women. IAS founders intended it to be inclusive, and women members were accepted from the earliest years. Goldman had directed excavations for the Fogg Museum. An intrepid traveler and meticulous excavator, her discoveries at Tarsus in Turkey are recorded in six volumes. She was photographed at right in 1912 in the Greek city of Halae, where she became fascinated by terra-cotta figures such as she holds below. Granddaughter of a cofounder of Goldman, Sachs, and Company, Goldman attended the American School of Classical Studies at Athens. She was a volunteer nurse in a Greek hospital during World War I and a sponsor of refugees during World War II. (Courtesy of Bryn Mawr College Library.)

Abraham Flexner staffed the School of Humanistic Studies within 18 months. Hetty Goldman joined Benjamin D. Meritt, Erwin Panofsky, Elias Lowe, and Ernst Herzfeld. Goldman lived close to the IAS campus on Newlin Road and spent summers in upstate New York, where this snapshot was taken in 1957. (Courtesy of the Historical Society of Princeton; Erich and Alice L. Kahler Collection.)

Ernst H. Kantorowicz (1895–1963) was appointed professor of medieval history in 1951. The German historian had studied at the University of Heidelberg and is remembered as an urbane and sophisticated epicure. He was also part of the circle drawn to Erich Kahler's Princeton home, One Evelyn Place. Kantorowicz (right) and Kahler are pictured at One Evelyn Place in 1958. (Courtesy of the Historical Society of Princeton; Erich and Alice L. Kahler Collection.)

Archaeologist and historian of Islamic and pre-Islamic studies, Ernst Herzfeld (1879–1948) joined the IAS faculty in 1936, following Hitler's expulsion of Jews from German universities. In 1919, when the photograph at right was taken, Herzfeld was associate professor at Berlin University. (Courtesy of Elizabeth Ettinghausen.)

Ernst Herzfeld was involved in excavations from Cilicia to Mesopotamia. He was fluent in Latin, Greek, Arabic, Persian, Turkish, and Hebrew, and his work resulted in landmark architectural surveys, including the monuments of the Tigris and Euphrates valleys. He published widely on Islamic architecture and ornament, including the palace at Persepolis, which was where the photograph at left was taken. (Courtesy of Oriental Institute of Chicago.)

American paleographer Elias Avery Lowe (1879–1969) was appointed professor in 1936. He spent a lifetime of scholarship in the field of *Codices Latini Antiquiores*, compiling a paleographical guide to Latin manuscripts, which he completed shortly before his death at age 90. He lived in Princeton with his wife, Helen Lowe-Porter, a celebrated translator of the works of Thomas Mann. (Courtesy of Louise Morse.)

American military historian Edward Earle (1894–1954) was appointed to become a professor in the new School of Economics and Politics in 1934. The school later merged with the School of Humanistic Studies to create the School of Historical Studies. Like most of his IAS colleagues during World War II, Earle was a consultant to various government departments. Gen. Carl A. Spaatz (left) congratulated Earle on his receipt of the Presidential Medal for Merit in 1946.

Five

JOHN VON NEUMANN
ELECTRONIC COMPUTER PROJECT

A practical hands-on research project involving engineers and vacuum tubes is far removed from the type of scholarship originally envisioned for the Institute for Advanced Study. Nevertheless, in 1946, director Frank Aydelotte and the board of trustees lent their support to John von Neumann's plans to build a high-speed digital computer at the institute.

Of all the stellar intellects drawn to the IAS in the early 1930s, none burned brighter than its youngest professor, John von Neumann. His lectures in Fine Hall drew large and lively audiences. At the afternoon teas that followed—a custom that continues to this day—von Neumann was the center of discussion. He tackled problems in pure mathematics, logic, and quantum theory and was forever moving into new territory. His interest in hydrodynamics would lead ultimately into high-speed computing.

During World War II, von Neumann's advice was sought by the U.S. military, and in 1943 by the Manhattan Project at Los Alamos. He contributed to the design of the ENIAC (Electrical Numerical Integrator and Computer), which, with the help of Oswald Veblen, was built to compute ballistics tables at the University of Pennsylvania.

Von Neumann recognized the potential for computers to go beyond their use as fast calculating devices. In late 1945, with funding from the IAS, the U.S. Atomic Energy Commission, the Office of Naval Research, and branches of the U.S. Army, the electronic computer project got underway in the basement of Fuld Hall. It was not long before it needed a building of its own.

As the "IAS Machine" developed, von Neumann distributed progress reports to scientists working on similar projects elsewhere. As a result, the basic architecture of the modern computer derives from its design, which influenced the first electronic computer marketed by International Business Machines—IBM 701.

In Princeton, it was used to solve fundamental problems in meteorology that made possible weather predictions of the sort taken for granted today. The "IAS Machine" functioned until 1958 and was retired to the Smithsonian Institution's National Museum of American History in 1960.

Born in Budapest, John von Neumann (1903–1957) was a child prodigy in languages and mathematics. By the time he was invited to lecture on quantum theory at Princeton University in 1930, his credentials were impeccable—doctorate at age 22, studies with Hilbert at Göttingen, and official appointments in Berlin and Hamburg. One of several Hungarian intellectuals to find a haven in Princeton, he became a professor at the Institute for Advanced Study in 1933. Like Veblen, von Neumann had been teaching at Princeton University, and the resulting furor forced Flexner to promise not to hire any more of the university's faculty. In addition to his numerous scientific contributions, von Neumann is remembered for his good-natured penchant for risqué stories and an ability to tell jokes in three languages simultaneously. This 1940s photograph shows von Neumann with his second wife, Klari Dan von Neumann, and their Irish setter, Inverse, in front of their Princeton home at 26 Westcott Road. This was the location of frequent parties that often went well into the night with cocktails, dancing, and dazzling conversation. (Courtesy of Marina von Neumann Whitman.)

This snapshot of John von Neumann walking with Marina, his 11-year-old daughter with his first wife, Mariette Kovesi, on a street in Santa Fe, New Mexico, was taken in 1946. Marina von Neumann grew up to become an influential economist. She teaches business administration and public policy at the University of Michigan and is a trustee of the Institute for Advanced Study. (Courtesy of Marina von Neumann Whitman.)

Described as "the 20th century's most scintillating intellect," John von Neumann was widely respected as a bright and gregarious colleague. His Hungarian-born wife, Klari Dan von Neumann, once quipped of her husband's photographic memory, "He will remember everything on a page in a book he read 15 minutes ago but he won't remember what he had for lunch."

A team of mathematicians, scientists, engineers, and technicians worked on the electronic computer project (ECP). On June 10, 1952, the team gathered for a formal dedication ceremony. Shown here are, from left to right, Ephraim Frei; Gerald Estrin; Lewis L. Strauss, IAS trustee; J. Robert Oppenheimer, IAS director; Richard Melville; Julian Bigelow, chief engineer from 1946 to 1951; Norman Emslie; James Pomerene, chief engineer from 1951 to 1956; Hewitt Crane; and John von Neumann.

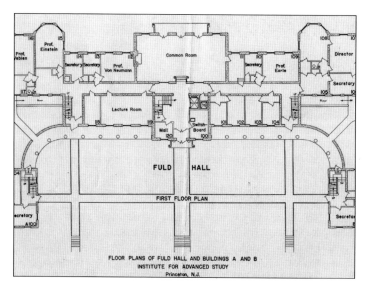

FLOOR PLANS OF FULD HALL AND BUILDINGS A AND B
INSTITUTE FOR ADVANCED STUDY
Princeton, N.J.

The electronic computer project brought an influx of practically skilled personnel, one of whom, Peter Panagos, prepared technical drawings of Fuld Hall. The first floor plan shows von Neumann's office next to that of Einstein. The ECP soon outgrew the Fuld Hall basement and moved to its own building on Olden Lane, which has since been renovated for use as a children's nursery school.

The ECP employed a large staff of engineers, programmers, machinists, draftsmen, secretaries, and data-entry clerks. The IAS machine's logical schema, known as "von Neumann architecture," would provide the model for subsequent stored-program computers. The machine cut its teeth on problems, which were supplied by mathematicians, biologists, astronomers, and meteorologists and painstakingly prepared by main programmer Hedvig Selberg, pictured in the middle in a striped dress, and her team of women programmers, using punch cards or paper tape. During the summer of 1951, the machine was used by scientists from Los Alamos for a complex thermonuclear calculation that ran for 24 hours at a time without interruption over a period of some 60 days. Among the throng pictured here are Julian Bigelow, Hewitt Crane, Gerald Estrin, Frank Fell, Ephraim Frei, Herman Goldstine (whose wife Adele also worked on the project), Richard Melville, Peter Panagos, and James Pomerene. (Photograph by Alan W. Richards.)

James Pomerene displays the Williams tube, a cathode ray tube used to store binary data, that replaced the "selectron" produced by RCA (Radio Corporation of America). The first random access digital storage device, which was made in England by F. C. Williams of Manchester, allowed instructions and data to be encoded into numeric form and stored, making it an important step in the evolution of computing. (Photograph by Alan W. Richards.)

Leon Harmon examines computer tape, the method by which instructions were fed into the IAS machine and results delivered. Harmon began his career as a wireman on the electronic computer project and went on to work on human perception, computer vision, and graphics for Bell Laboratories. (Photograph by Alan W. Richards.)

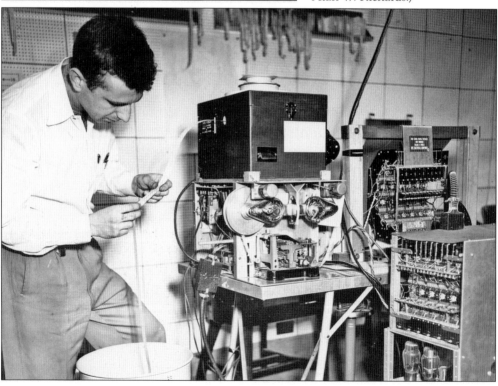

In the early 1950s, the IAS became the leading center for new ways of thinking about weather and climate. Meteorologists Jule Charney, Joseph Smagorinsky, Norman Phillips, and others worked on climate modeling and numerical weather prediction. Smagorinsky became the first director of the National Oceanic and Atmospheric Administration's Geophysical Fluid Dynamics Laboratory. Phillips was the first to show that weather prediction with numerical models was feasible. For their work, which changed the art of weather watching into a science, Phillips and Smagorinsky received the Franklin Medal in Earth Sciences in 2003. The 1952 photograph above shows, from left to right, Smagorinsky, Phillips, Herman H. Goldstine, and Gerald Estrin. The photograph below shows J. Robert Oppenheimer (left) with John von Neumann in front of the IAS machine. (Photographs by Alan W. Richards.)

In 1954, John von Neumann was asked to serve as one of five atomic energy commissioners. A year later, he was diagnosed with bone cancer. He was already fatally ill when he received the Presidential Medal of Freedom from Pres. Dwight D. Eisenhower at the White House in 1956. He died at the age of 53 on February 8, 1957, and is buried in the Princeton cemetery. His wife, Klari von Neumann, who is thought to have committed suicide by drowning in 1963, is buried beside him. Von Neumann's ideas on the computer and the human brain, developed for a series of lectures to be presented at Yale University, remained in unfinished manuscript form until published by Yale University Press in 1958. (Courtesy of Marina von Neumann Whitman.)

Six

J. ROBERT OPPENHEIMER
CHARISMATIC LEADERSHIP

When theoretical physicist J. Robert Oppenheimer succeeded Frank Aydelotte to become the Institute for Advanced Study's third director in 1947, he also joined the faculty, making him the first director to do so. Oppenheimer had led the Manhattan Project at the Los Alamos National Laboratory during World War II and was world-renowned as "father of the atomic bomb." Oppenheimer would direct the IAS for almost two decades.

During that time, the institute changed enormously, as did Oppenheimer's public life. After the war, as chairman of the advisory committee of the Atomic Energy Commission (AEC), he opposed the development of the hydrogen bomb. Oppenheimer's stance was controversial at a time when Sen. Joseph McCarthy's fears of Communist subversion were gripping the nation. Oppenheimer lost his position with the AEC and his security clearance. He did not, however, lose directorship of the Institute for Advanced Study, where he was a charismatic and energetic leader.

Oppenheimer consolidated the School of Economics and Politics and the School of Humanistic Studies that Flexner had established in the first half of the 1930s into a new School of Historical Studies. He appointed new faculty and expanded the campus with new office buildings, as well as a Marcel Breuer–designed "village" for members who were still being housed in drafty war-surplus barracks. He commissioned Wallace K. Harrison to design a new Historical Studies-Social Science Library. He also recognized the fact that many of the professors and members in the School of Mathematics were actually physicists and formed a new School of Natural Sciences.

Under Oppenheimer, the IAS became the world's center for high-energy physics and field theory, attracting young particle physicists while retaining solid contact with established figures like Niels Bohr, Wolfgang Pauli, and Paul Dirac. He appointed Abraham Pais, Freeman Dyson, Tsung-Dao Lee, and Chen Ning Yang to the faculty. Together, Lee and Yang would receive the Nobel Prize in Physics in 1957. Oppenheimer directed the Institute for Advanced Study until 1966 and died of throat cancer the following year.

J. Robert Oppenheimer (1904–1967) was born into a well-to-do Manhattan family. His father, Julius Oppenheimer, came to the United States from Germany to work with relatives in the textile trade; his mother, Ella Friedman, was an accomplished painter who had studied in Paris and taught at Barnard College. Oppenheimer attended the Ethical Culture School in New York City and graduated *summa cum laude* from Harvard University. After studying in Europe at Cambridge and Göttingen, he accepted two teaching positions simultaneously, at the California Institute of Technology and the University of California, Berkeley. Oppenheimer made such an impression as a teacher that his students adopted his mannerisms and style of speaking. Institute trustee Lewis L. Strauss was chairman of the search committee for Aydelotte's successor, which offered Oppenheimer the directorship in 1947. In 1954, Strauss would be the driving force behind the hearings that revoked Oppenheimer's security clearance. Pres. Lyndon B. Johnson, following the wishes of his predecessor John F. Kennedy, awarded Oppenheimer the $50,000 Fermi Prize in 1963, by way of a national apology. (Photograph by Ulli Steltzer.)

As IAS director, Oppenheimer and his family lived at Olden Manor, now known as Olden Farm, at 97 Olden Lane. The house dates to the late 17th century and is named for one of New Jersey's most prominent Quaker families, which included Charles Smith Olden, governor of New Jersey from 1860 to 1863. Oppenheimer had a greenhouse added, which has since been removed, so that his wife could grow orchids. This June 1949 photograph shows Oppenheimer with his wife, Katherine ("Kitty") Puening Harrison, whom he had married in 1940, their son Peter, born in 1941, and their daughter Katherine ("Toni"), born in 1944. The couple entertained frequently at Olden Manor, which continues to be a residence for IAS directors. (Right, by Alfred Eisenstaedt; courtesy of Getty Images.)

Dutch particle physicist Abraham Pais (1918–2000) observed that Oppenheimer brought to the IAS his "outstanding talent for assembling the right people and stimulating them to great effort." Members could discuss their ideas with the likes of Einstein, Wolfgang Pauli, Niels Bohr, and Paul Dirac and brainstorm their latest theories in lively seminars with their peers. In 1947, when these photographs were taken, Pais (at right) was one of a slew of youthful scientists—Freeman Dyson arrived the following year—who gravitated toward the IAS. Oppenheimer appointed Pais to the faculty in 1950 and Dyson in 1953. The 1933 Nobel Laureate Dirac, a founder of quantum mechanics, was a frequent visitor. Below, sessions involving Oppenheimer (left), Dirac (center), and Pais began in seminar and were then continued over afternoon tea. (Photographs by Alfred Eisenstaedt; courtesy of Getty Images.)

Until it was replaced by a spacious dining hall in 1971, a no-frills cafeteria on the fourth floor of Fuld Hall served the IAS community. During the summer months, the cafeteria was unbearably hot. It was often so packed that one had to eat quickly to make room for others. Here Dirac (center) converses with Pais (right) and others. (Photograph by Alfred Eisenstaedt.)

This 1954 photograph captures four Nobel Prize winners associated with the IAS, and they are, from left to right, Niels H. D. Bohr (1885–1962), 1922 laureate; Arthur H. Compton (1892–1962), 1927 laureate; Albert Einstein (1879–1955), 1921 laureate; and Isador Isaac Rabi (1898–1988), 1944 laureate. Towards the latter part of his life, Einstein's dogged search for a unified field theory was regarded as "old physics" by the younger set at the institute.

Theodore H. Berlin (1917–1961) took sabbatical leave from Johns Hopkins University to come to the institute for the academic year 1952–1953. Berlin's research was published in a collaborative paper with his student Louis Witten, whose son Edward Witten would later become an IAS professor and superstring theorist. (Courtesy of Dennis Berlin.)

In his biography *J. Robert Oppenheimer: A Life*, Abraham Pais (left) described Japanese physicist Hideki Yukawa (1907–1981) as soft-spoken and shy. Yukawa, accompanied by his wife, Sumiko, came to Princeton as an IAS member for the academic year 1948–1949. A year later, he was Japan's first recipient of the Nobel Prize for his influential work in particle physics.

In 1957, Tsung-Dao Lee (left) and Chen Ning ("Frank") Yang received the Nobel Prize for their work on parity violation. Yang first came to the IAS as a member in 1949 and became a professor in 1955, staying until 1966. He has described his 17 years there as "the period when I did my best research work." Lee was on the faculty from 1960 until 1962.

French-born André Weil (1906–1998) was among the mathematicians appointed by Oppenheimer in the 1950s. Brother of the philosopher Simone Weil, he is remembered for work in number theory and algebraic geometry. In 1994, when this photograph was taken, Weil received the Inamori Foundation's Kyoto Prize, often called Japan's Nobel Prize. (Photograph by Randall Hagadorn.)

Together with his wife, Louise, Marston Morse was host to several gatherings of mathematicians a year over several decades. The Morse residence on Battle Road was second home to an international set visiting Princeton University and the IAS. In 1960, Louise Morse photographed, from left to right, (seated) French mathematician Jean Leray, a frequent visitor during the 1950s, and Marston Morse; (standing) logician Richard C. Jeffrey, unidentified (possibly Irwin E. Vas), and French mathematician Paul Malliavin. (Courtesy of Louise Morse.)

Born in India, Harish-Chandra (1923–1983) was a student at Cambridge University in England when he pointed out a mistake in a lecture given by Wolfgang Pauli. At Cambridge, Harish-Chandra's doctoral thesis adviser was Paul Dirac, as whose assistant he visited the IAS in 1947. He was appointed professor in 1963. Harish-Chandra switched from theoretical physics to fundamental studies in mathematics, and the Harish-Chandra Research Institute in Allahabad, India, is named in his honor. (Courtesy of Lalitha (Lily) Harish-Chandra.)

The mathematics community gathered for a conference at the IAS in September 1957. Participants who sat for this photograph behind Fuld Hall included, in the second row, Marston Morse, seventh from the right; Arne Beurling, third from the right; and Atle Selberg, first from the left. If the Nobel Prize were offered for mathematics—it is said that its founder specifically excluded this subject because of a personal animosity to one particular mathematician—the number of IAS laureates would be impressive. The Nobel equivalent in mathematics is the Fields Medal. To date, almost 75 percent of all Fields Medal recipients have been affiliated with the Institute for Advanced Study.

Colleagues of Marston Morse attended a symposium in April 1963, following his retirement the year before. Morse, who was born in Waterville, Maine, was celebrated for defining a new branch of mathematics known as Morse Theory, for his emphasis on the affinity between mathematics and art, and for his distinguished efforts during two world wars. In World War I, Morse served with the American Expeditionary Force and received the *Legion d'honneur* and the *Croix de Guerre* from the French government. For his contributions during World War II, he received the Presidential Certificate of Merit in 1947. Shown from left to right are (first row) Shing-Shen Chern, R. G. Pohrer, Atle Selberg, Marston Morse, Walter Leighton, Morris Hirsch, Stewart S. Cairns, and Hassler Whitney; (second row) Raoul Bott, Barry C. Mazur, Gustav A. Hedlund, Theodore T. Frankel, Stephen Smale, Nicolaas H. Kuiper, J. Frank Adams, William Browder, John W. Milnor, and Michel A. Kervaire. (Photograph by Alan W. Richards.)

While their home on Maxwell Lane was being built in 1952, Norwegian number theorist Atle Selberg (1917–2007), his wife Hedvig Selberg, and daughter Ingrid Selberg were living in what was known as the IAS housing project. After being a member from 1947 to 1948 and a permanent member from 1949 to 1951, Selberg joined the faculty in 1951. (Courtesy of Ingrid Selberg.)

In 1948, Deane Montgomery (1909–1992) was one of the first permanent members. He became a professor in 1951 and subsequently contributed to the resolution of Hilbert's fifth problem, one of 23 unsolved problems in mathematics listed at the turn of the 20th century. This 1977 photograph of Montgomery (right), with Princeton resident David Duvivier, was taken at the home of Marston and Louise Morse. (Courtesy of Louise Morse.)

Swiss mathematician Armand Borel (1923–2003) came to the Institute for Advanced Study with his wife, Gabrielle ("Gaby") Aline Borel, for two years in 1952 and joined the faculty in 1957. A devotee of jazz and classical music, Borel directed the first IAS concert series from 1985 through 1992, organizing concerts by Indian flutist Shashank Subramanyam, the New York Camerata, and Art Farmer's Jazz Quartet, among others. (Photograph by Gary Alvin.)

The 1966 Fields Medalist Michael Atiyah was on the IAS faculty from 1969 until 1972 before returning to his native England and pursuing a distinguished career that included a knighthood and presidency of the Royal Society. Atiyah (second from left) and his wife, Lily Brown Atiyah, were photographed at the home of Marston and Louise Morse with, from left to right, Jean-Pierre Serre of France and Shunich Tanaka of Japan. (Courtesy of Louise Morse.)

Tullio Regge was one of a number of physicists working within the School of Mathematics by the mid-1960s. The physicists formed a working group within the school until 1966, when Oppenheimer formed the School of Natural Sciences. In 1979, Regge received the IAS Einstein Award. He returned to his native Italy that year and was elected to the European parliament in 1989. (Photograph by Anna-Greta Larsson Wightman.)

Archaeologist Homer A. Thompson (1906–2000) was appointed in 1947. Thompson's life's work was excavating the Agora, the ancient Athenian marketplace, which he rediscovered by following in the footsteps of the Greek traveler and geographer Pausanias. In Athens, Thompson worked with the American School of Archaeology, which for a time had a presence at the IAS. (Photograph by Herman Landshoff.)

In 1948, Harold F. Cherniss (1904–1987) was Oppenheimer's first faculty appointment. Born in Missouri and educated at the University of California, Berkeley, Cherniss (right) was the country's foremost expert on Plato and Aristotle. During World War II, he was assigned to a British intelligence unit in Europe. He was photographed in 1986 with former member Klaus H. Oehler. (Photograph by Elisabeth Oehler.)

Marshall Clagett (1916–2005) was appointed to the faculty in 1964, having been a member on two occasions. A historian of science, Clagett is celebrated for his landmark five-volume *Archimedes in the Middle Ages*, published between 1964 and 1984. This photograph dates to 1996. At the time of his death, Clagett was working on the fourth and final volume of his *Ancient Egyptian Science*. (Photograph by Randall Hagadorn.)

Born in Cincinnati, Ohio, Millard Meiss (1904–1975) came to the IAS in 1958 from Harvard University, where he was curator of paintings at the Fogg Art Museum. As chairman of the American rescue effort to save Italian art threatened by the 1966 Florence floods, Meiss helped restore damaged frescoes and buildings as recorded in his *Great Age of Fresco: Discoveries, Recoveries, and Survivals*, published in 1970.

The appointment of Oppenheimer's friend George F. Kennan (1904–2005) to the School of Historical Studies in 1956 was controversial with the existing IAS faculty, primarily because Kennan was a career diplomat rather than an academic scholar. But Oppenheimer, who admired Kennan's intellect, prevailed, and Kennan embarked upon a new career, publishing prolifically on the history of Russia and the Soviet Union. (Photograph by Constance Goodman.)

Seven

CARL KAYSEN
ADDING A FOURTH SCHOOL

Carl Kaysen took up leadership of the Institute for Advanced Study during a period of social upheaval. Flexner's hurried attempt to bring in economists who would address some of the world's problems had foundered. Oppenheimer had folded the School of Economics and Politics into a School of Historical Studies. Kaysen's appointment signaled renewed interest in topics relevant to contemporary society. He came in with a mandate for change.

Kaysen was a Harvard-trained economist with a career in public service that included deputy special assistant for national security affairs for Pres. John F. Kennedy from 1961 to 1963. Described as one of the president's "wisest counselors," Kaysen had negotiated the 1963 Partial Test Ban Treaty between the United States, the Soviet Union, and the United Kingdom.

As IAS director from 1966 to 1976, Kaysen introduced a program for social change that ushered in a fourth school. He invited the influential anthropologist Clifford Geertz to build up the new School of Social Science.

From time to time, the IAS has excited media curiosity and has been variously described as "a temple for promising intellectuals," "a genius club," and "the penthouse of the Ivory Tower." In 1973, however, media attention catapulted the quiet campus into the limelight in a less dignified manner.

When Kaysen and Geertz tried to expand the fledgling School of Social Science, they faced opposition from the existing faculty, and the "revolt" made headlines. In March 1973, the front page of the *New York Times* announced "Dispute Splits Advanced Study Institute." Sensational reports followed. An article in the *Atlantic Monthly*, "Bad Days on Mount Olympus," described the internal dispute as a "shoot-out" and Kaysen as the "arch-enemy of the geniuses."

Ultimately, the School of Social Science weathered the storm. Under Kaysen, the IAS campus expanded to include an office building for historians and social scientists and a much-needed dining hall to replace the utilitarian cafeteria. With the new dining hall and a world-class chef to run it, the IAS became famous for its cuisine. A culture of insularity would begin to give way to one of collegiality centered on a gathering place for a true community of scholars.

Born in Philadelphia, Carl Kaysen (1920–2010) earned his bachelor's degree from the University of Pennsylvania in 1940 and then went to work for the Office of Strategic Services. After World War II, he received a master's and doctoral degrees from Harvard University and was appointed a professor of economics there in 1957. At the Institute for Advanced Study, Kaysen encountered a culture of old world formality and a dearth of communication among a faculty that divided according to disciplines. He set about promoting social discourse through a series of faculty lectures. With his wife, Annette Kaysen, he invited professors and members to once-a-month dinners. In addition to shaping the School of Social Science, Kaysen raised money to fund it and to create new buildings for the campus. The school is his enduring legacy. (Courtesy of Allen H. Kassof.)

In 1970, anthropologist Clifford Geertz (1926–2006) took up the challenge of building a new School of Social Science. Born in San Francisco, Geertz had served in the U.S. Navy during World War II and then used the GI Bill to gain a bachelor's degree from Antioch College and a doctorate from Harvard University. Renowned for extensive fieldwork in Indonesia and Morocco, Geertz was teaching at the University of Chicago, when Kaysen asked him to come to the IAS. His "interpretive" approach to social science would define the new school, stressing the interplay of culture and traditional structures for an understanding of social change. Geertz's books have been translated into over 20 languages. His work on religion (especially Islam), economic development, and traditional village and family life influences anthropologists, geographers, ecologists, political scientists, and historians throughout the world. (Photograph by Randall Hagadorn.)

In 1974, the development economist Albert O. Hirschman (above, left) joined the School of Social Science. Born in Berlin and educated in Paris, London, and Trieste, Hirschman had helped the American journalist Varian Fry rescue artists and intellectuals from the Nazis during World War II. At the IAS, he was a major voice for change in Latin America. Institute driver Gary Alvin photographed Hirschman and his wife, Sarah, founder of the literature discussion program People and Stories/Gente y Cuentos, with their two French grandchildren, Lara and Grégoire Salomon, at Newark airport in the 1980s. Alvin also photographed political and moral philosopher Michael Walzer (left), as he traveled from Princeton Junction. Like Geertz and Hirschman, Walzer combines theoretical issues with concrete political activity. He joined the School of Social Science in 1980. (Photographs by Gary Alvin.)

In 1985, Joan Wallach Scott became the Institute for Advanced Study's first woman professor since Hetty Goldman. A 1962 graduate of Brandeis University, Scott received her doctorate from the University of Wisconsin, Madison, in 1969. She has since defined the field of gender history and challenged the foundations of conventional historical practice, including the nature of historical evidence and historical experience and the role of narrative in the writing of history. (Photograph by Cliff Moore.)

In 2007, Eric Maskin, who was appointed in 2000 as the first Albert O. Hirschman professor, received the Nobel Prize for his work in economics. He lives on Mercer Street in the former Einstein residence. He is the third IAS professor to do so and, coincidentally, the third Nobel Laureate to live at that address. (Courtesy of Eric Maskin.)

In addition to social science, Kaysen also made appointments in the three other schools. Kenneth M. Setton (1915–1995), an authority on the medieval Crusades, joined the School of Historical Studies in 1968. Born in New Bedford, Massachusetts, Setton graduated from Boston University before gaining his doctorate from Columbia University. He devoted almost two decades of scholarship to his four-volume classic study *The Papacy and the Levant, 1204–1571.* (Photograph by Ulli Steltzer.)

Known as the "Pope of plasma physics," Marshall N. Rosenbluth (1927–2003) was appointed by Kaysen in 1967. Having studied with Edward Teller in Chicago and worked on the hydrogen bomb, Rosenbluth (left) mentored a generation of plasma theorists, including James W. Van Dam (right), before leaving the institute in 1982. He directed fusion studies at the University of Texas at Austin. (Photograph by Gary Alvin.)

Stephen L. Adler (right) and Roger Dashen (below) were both promising young, high-energy physicists when Kaysen appointed them long-term members in 1966 and then full professors in 1969. Dashen (1938–1995) resigned in 1987 to take up a position at the University of California, San Diego. Adler, now professor emeritus, received the Dirac Medal of the International Center for Theoretical Physics for "significant contributions to theoretical physics" in 1998. In 1979, when the photograph at right was taken, Adler was participating in a discussion of "Einstein and the Physics of the Future," which was part of the centennial symposium to celebrate the achievements of Albert Einstein.

Astrophysicist John N. Bahcall (1934–2005) joined the School of Natural Sciences in 1971. By his own account, his interest in physics developed simply because he had been required to take a science class at the University of California, Berkeley. At high school in Shreveport, Louisiana, he had focused on tennis and debating, becoming a state tennis champion and a national debate team champion. After graduating with a bachelor's degree in physics from Berkeley in 1956, he obtained a master's from the University of Chicago in 1957 and a doctorate from Harvard University in 1961. At the IAS, Bahcall's enthusiasm for subjects such as solar neutrinos, black holes, and dark matter inspired researchers throughout the world. He published over 600 scientific papers and five books. His contributions to the Hubble telescope were acknowledged in 1992, when he received NASA's distinguished public service medal. He is fondly remembered walking at the center of a group of young scientists every day from the science building to the dining hall. (Photograph by Randall Hagadorn.)

Eight

A WORLD APART
CAMPUS LIFE

All work and no play makes for a very dull day. When the Institute for Advanced Study moved to Fuld Hall in 1939, the main social activity was the daily ritual of afternoon tea. Professors organized bowling matches on the lawn in front of Fuld Hall and installed a radio in the common room for occasional Saturday evening dances for the younger members.

World War II was an extraordinary time of activity with professors contributing to the war effort in their own ways. But after the war, the IAS, like the rest of the nation, returned to its own business. Scholarly activities broadened, the campus expanded physically, and in 1947, several faculty wives formed Crossroads Nursery School for the young children of professors and members living in the recently erected housing complex of old army barracks.

Although Oppenheimer famously said that his role was to make sure that scholars found no distractions and therefore had no excuses for not focusing on their research, successive administrations have looked for ways not only to improve resources that make a stay at the IAS as uncomplicated and comfortable as possible for members and their families but also to increase opportunities for cultural and social activities.

Since Carl Kaysen, the IAS has been directed by Harry Woolf, Marvin Goldberger, Phillip Griffiths, and Peter Goddard. As the institute expanded from one to four schools, campus life and social activities also expanded. The Association of Members of the Institute for Advanced Study (AMIAS) was founded in 1974. It was developed to promote the growth of the institute and to forge a connection to the larger community. The Friends of the Institute for Advanced Study was founded by a small circle of Princeton supporters in 1980.

Besides faculty and members, the IAS also comprises a staff that facilitates the daily operation of the campus and celebrations of milestones such as the centennial symposium for Albert Einstein in 1979 and the 50th and 75th anniversaries of the institute in 1980 and 2005.

The tradition of tea at 3:00 p.m. dates to the 1930s and to Elizabeth Veblen, wife of Oswald Veblen. In the photograph above, Elizabeth Veblen pours a "cuppa" for an unidentified scholar during the summer in Maine. At that time, it was customary for faculty wives to lend a hand in preparing and pouring afternoon tea. In 1963, during a symposium that honored mathematician Marston Morse (below), participants took a break for tea, which was served in fine china on a lace tablecloth. The otherwise Spartan-looking common room looks different today. The linoleum tiles have been replaced by hardwood flooring and oriental rugs. The grandfather clock remains in good working order. (Below, photograph by Alan W. Richards.)

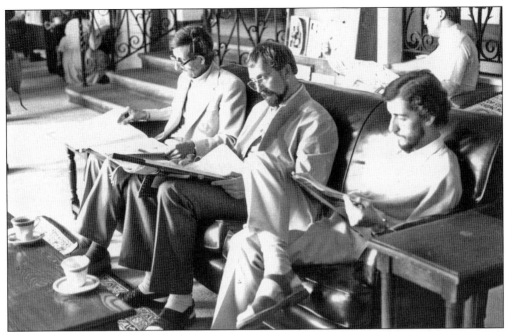

The common room is also the place to catch up with the daily news. Physicist and mathematician Freeman J. Dyson (left) was photographed with unidentified members in 1985. Dyson came to the IAS from his native England in 1948 and was a member in mathematics before being appointed professor in 1953. When the School of Natural Sciences formed in 1966, Dyson was one of its first professors. (Photograph by Linda Troeller.)

Paper cups have replaced china cups and saucers in this more recent 2004 photograph, in which Oleg Grabar (center) converses with fellow professor emeritus Morton White (left) and unidentified visitors. An expert in Islamic art and culture, Grabar joined the faculty in 1990. White was appointed in 1970 as the first professor to concentrate on both the history of American thought and systematic philosophy. (Photograph by David Graham.)

As the IAS grew during the 1940s, so did the need for housing and other resources. Crossroads Nursery School was formed to serve the needs of IAS parents in 1947. It was housed in a World War II vintage clapboard building at the bottom of Olden Lane, where these photographs were taken. In this 1955 snapshot (at left), teachers serve up milk for youngsters. In the 1952 photograph (below), children celebrate a traditional Halloween. Those pictured include, from left to right, (first row) unidentified, Tamara ?, Katarina Hafaeli (in Arab headdress), Rebecca Kaplan (in front of clown, wearing a crown), Esther Dyson, Nora Charney, Ingrid Selberg, and Alice Bigelow (in white). Since these days, the nursery school has moved to 225 Olden Lane, which formerly housed the IAS electronic computer project. (Left, courtesy of Ingrid Selberg; below, photograph by Verena Huber-Dyson.)

Ingrid Selberg, shown at right with her parents Atle and Hedvig Selberg and her brother Lars, was among the first group of IAS children to attend Crossroads Nursery School in the early 1950s. The Selbergs built a home nearby on Maxwell Lane. Atle Selberg had come to the institute in 1947 and won the Fields Medal in 1950. In 1987, on his 70th birthday, Selberg was honored as Commander of the Order of St. Olav with Star by the King of Norway in Oslo. The formal dinner was attended by Ingrid Selberg, her parents, and brother Lars. (Courtesy of Ingrid Selberg.)

Historian of science Harry Woolf succeeded Carl Kaysen and served as director from 1976 to 1987. During Woolf's tenure, the IAS modernized with computer technology, formed an Association of Members of the Institute for Advanced Study (AMIAS), and a formal group of Friends of the Institute for Advanced Study. In 1979, on the centennial of the birth of Albert Einstein on March 14, 1879, Woolf directed a symposium celebrating Einstein's achievements.

The Einstein Centennial Symposium attracted scientists and historians from around the world, including Hans Bethe, P. A. M. Dirac, Stephen W. Hawking, Martin J. Rees, and Roger Penrose. Physicist Marvin L. Goldberger, pictured left, moderated a session on "Einstein and the Physics of the Future," in which Freeman Dyson also participated. Goldberger would succeed Harry Woolf in 1987 and serve as the sixth director until 1991.

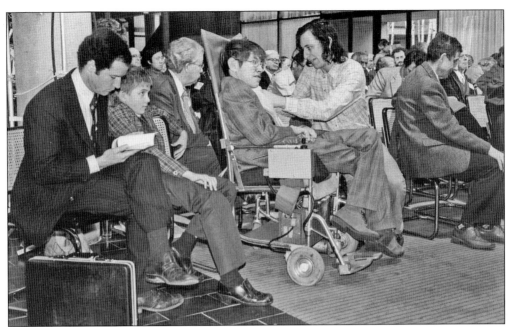

British theoretical physicist and cosmologist Stephen W. Hawking was already confined to a wheelchair by the incurable motor neuron disease ALS (amyotrophic lateral sclerosis) when he joined the international group of scientists attending the Einstein Centennial in 1979. That same year, Hawking followed his illustrious predecessor Isaac Newton in being appointed Lucasian Professor of Mathematics at Cambridge University. P. A. M. Dirac, who held the Lucasian chair from 1932 to 1969, was also at the event.

British cosmologist and astrophysicist Martin J. Rees spoke on "The Size and Shape of the Universe," as part of the institute's 1979 Einstein Centennial Symposium. Rees, who was appointed Britain's Astronomer Royal in 1998, was a member in the School of Natural Sciences from 1969 to 1970 and on several other subsequent occasions. He is currently on the IAS board of trustees.

The year 1979 saw the death of Minot C. ("Mike") Morgan Jr., general manager and comptroller of the IAS for almost 25 years. Carl Kaysen commended Morgan's contributions to the "effective functioning" of the institute. Morgan's diverse staff, said Kaysen, was "characterized by low turnover and long service." Morgan was mayor of Princeton borough from 1946 to 1948.

In 1980, the Institute for Advanced Study marked its 50th anniversary with a dinner. Trustee emeritus Harold F. Linder, left, celebrated with Mrs. J. Richardson ("Bunny") Dilworth (Elizabeth McKay Cushing) and trustee Donald B. Straus. (Photograph by Alan Hess.)

J. Richardson Dilworth (1916–1997) was a trustee of the Institute for Advanced Study for 34 years and is remembered as a wise counselor who participated in the long and ultimately successful negotiations to protect the IAS woods and farmlands. Elected to the board in 1964, Dilworth subsequently served as its president, vice chairman, and chairman. A graduate of Yale College and Yale Law School, Dilworth began in business with the banking firm of Kuhn, Loeb, and Company and was senior financial adviser to the Rockefeller family for 23 years. He was chair of Rockefeller Center and the Metropolitan Museum of Art. The "Dilworth Room" and a fellowship in the School of Historical Studies are named in his memory. (Photograph by Mary Cross.)

To cater its Einstein Centennial event in 1979, the IAS hired an outstanding chef, who would stay on and ultimately transform the dining hall into Princeton's best culinary experience. Franz Moehn, pictured in the center, with staff Domingo Tanglaw and Barbara Paul, has been described as one of the institute's most revered figures. Moehn introduced fine wine and fine cuisine.

The tradition of fine dining at the Institute for Advanced Study that began with Chef Moehn continues today with Chef Michel Reymond, who was hired in 1999. In 2007, Chef Reymond lit the candles for mathematician Atle Selberg's 90th birthday celebration, while the mathematician's wife Betty Selberg looked on. (Photograph by Dinah Kazakoff.)

Led by John E. Ozwirk Sr. (left), members of the Institute for Advanced Study's maintenance crew John E. Ozwirk Jr. (center) and Michelino ("Mike") Antenucci make for the dining hall during the institute's 50th anniversary in 1980, while Trenton resident Banks A. Hairston Sr. (at right) does duty in the kitchen. Hairston worked at the IAS from 1970 to 1987. (Photographs by Gary Alvin.)

James H. Barbour Jr. was on the IAS staff for 25 years before retiring in May 2005. He had worked at Princeton University before accepting an offer from IAS associate director Allen Rowe. Barbour, who was responsible for buildings and grounds, is credited with creating an environment that facilitated the achievements of faculty and members. He was universally respected for his integrity and determination. (Photograph by Gary Alvin.)

In November 1981, a sculpture by minimalist artist Tony Smith was unveiled next to the IAS pond. Titled *New Piece*, the sculpture was a gift from the artist to commemorate Einstein's birth in 1879. Smith, who was born in South Orange in 1912, died in 1980, prior to the installation. A retrospective of his work was held at the Museum of Modern Art in New York in 1998. (Photograph by W. Brinton Whitall.)

Institute employees photographed in the print shop in Fuld Hall in May 1981 are, from left to right, James ("Jim") Bess, Glenda Whitmore Johnson, and James ("Jim") Farrior. (Photograph by Gary Alvin.)

Edith Kirsch (left), an art historian working as assistant to Prof. Millard Meiss in the School of Historical Studies, and Caroline Underwood (right), administrator for the School of Mathematics, celebrated with Elizabeth Horton, secretary of the School of Historical Studies, on the occasion of Horton's retirement in 1981. (Photograph by Gary Alvin.)

Gary Alvin (above) stands with the IAS vehicle he used to drive faculty and members in the late 1970s and early 1980s. Alvin worked in the mailroom from 1964 until 2003 and was a keen photographer. In this 1981 photograph (below), mathematician Robert P. Langlands and his wife, Charlotte Langlands, set off on summer vacation. Langlands, who joined the faculty in 1972 and was Hermann Weyl Professor in the School of Mathematics from 1993 until he became professor emeritus in 2007, is regarded as one of the most significant mathematicians of the 20th century. He inherited Einstein's office. Charlotte Langlands is a noted sculptor who teaches classes for members of the IAS community. (Courtesy of Gary Alvin.)

Freeman Dyson, Imme Dyson, and two of their daughters set off on a journey in 1981. From left to right are unidentified, Freeman Dyson, Imme Dyson, Mia Dyson, and Rebecca Dyson. Eighteen-year-old Dorothy Dyson, pictured below with Gary Alvin, found work in the IAS mailroom in 1978, the summer after her freshman year at University of California, Davis (UCD). (Courtesy of Gary Alvin.)

Marvin Goldberger and his wife, Mildred Goldberger, were photographed arriving in the dining hall in 1990. During Goldberger's time as director, which was from 1987 to 1991, the popularity of public lectures given by faculty, as well as a desire for a venue for concerts, prompted plans for an auditorium on campus. A new building for the School of Mathematics was also considered. (Photograph by Gary Alvin.)

Astrophysicist Piet Hut was appointed to the School of Natural Sciences in 1985. Since then, Hut has forged a unique place for himself as a professor in his own program in interdisciplinary studies, which brings together aspects of physics, astrophysics, geology, artificial intelligence, cognitive psychology, philosophy, and computer science. (Photograph by Pryde Brown.)

String theorist Edward Witten became a professor in the School of Natural Sciences in 1987, after leading a program that established increased contact between mathematicians and physicists. Witten, who had received his doctorate from Princeton University in 1976 and taught there from 1980 to 1987, won a Fields Medal in 1990. He is considered the man most likely to succeed in achieving Einstein's goal of finding the "theory of everything." (Photograph by Randall Hagadorn.)

The 2004 Nobel Laureate Frank Wilczek was a theoretical physicist in the School of Natural Sciences from 1989 until 2000, when he joined the Massachusetts Institute of Technology. Wilczek is one of the three Nobel prizewinners who have lived at Einstein's old address of 112 Mercer Street, and the others are, of course, Einstein and the economist Eric Maskin. (Photograph by Randall Hagadorn.)

Princeton resident Laura Hill, who runs a childcare center on Spruce Street, said hello to actor Walter Matthau, while walking with her charges in Princeton, during a lull in filming for the 1994 movie *IQ*. A passerby took this snapshot of Hill with Emma Karp, Daniel Zack, and Matthau as Einstein. While several documentaries have been filmed at the IAS and at the Einstein residence on Mercer Street, *IQ*, starring Meg Ryan and Tim Robbins, is the only full-length feature to be shot there. The romantic comedy is a work of pure fiction with barely a nod in the direction of fact. Kurt Gödel, who was shy in real life, is depicted as positively exuberant in the movie. (Left, courtesy of Robert Landau.)

Mathematician Phillip A. Griffiths (right) became the seventh director of the IAS in 1991 and served until 2003. In 1997, Griffiths attended a board meeting with trustee Michael Bloomberg, later mayor of New York City. One topic under discussion was the renovation of member housing, which was completed in 2001. (Photograph by Randall Hagadorn.)

The IAS welcomed its first artist-in-residence in 1994, pianist Robert Taub (right). Seen here with Princeton philanthropist and IAS supporter William B. Scheide, Taub performed the complete cycle of Beethoven sonatas to an appreciative Princeton audience. Since Taub, composers Martin Butler, Jon Magnussen, and Paul Moravec and composer and clarinetist Derek Bermel have been artists-in-residence. Their public performances are part of an annual series of lectures and concerts that are presented free to the Princeton community. (Courtesy of Public Affairs, Institute for Advanced Study.)

Christian Habicht and his wife, Freia Habicht, attend the IAS midwinter party on February 13, 1988. Habicht, who was appointed professor in the School of Historical Studies in 1973, continues the long tradition of scholarship in Greek history, which has been a part of the IAS focus since its earliest days. (Photograph by Randall Hagadorn.)

Princeton residents and Friends of the Institute for Advanced Study, Allen H. Kassof (second from left) with his wife, Arianne Kassof (right), attend a social event with Portuguese diplomat José Cutileiro, the George F. Kennan professor in the School of Historical Studies from 2001 to 2004, and his wife, Myriam Sochacki. Throughout the year, Friends of the Institute for Advanced Study participate in the intellectual and cultural life of the IAS through lectures, concerts, and social gatherings.

Princeton residents Karl Morrison, medieval historian and former member, and Lloyd Moote, historian and member of the Friends of the Institute for Advanced Study, examine volumes in a special 75th anniversary exhibition. A rare book collection, the gift of Lessing J. Rosenwald, included first editions of important works in mathematics, astronomy, physics, and life sciences and is part of the Historical Studies-Social Science Library. (Photograph by Cliff Moore.)

John Forbes Nash Jr. was among the former members to attend the 75th anniversary celebrations in 2005. The 1994 Nobel Laureate is the subject of a biography by Sylvia Nasar, *A Beautiful Mind,* subsequently made into a film starring Russell Crowe. The film focuses on Nash's mathematical genius and struggle with schizophrenia. Nasar worked on her book when she was a director's visitor at the Institute for Advanced Study. (Courtesy of Public Affairs, Institute for Advanced Study.)

On May 20, 2005, the IAS celebrated its Founders Day with the dedication of a new sculpture by artist Elyn Zimmerman, the widow of the late IAS professor Kirk Varnedoe. The sculpture was the gift of trustee Robert B. Menschel. Pictured from left to right, the event was attended by Menschel, Princeton Township mayor Phyllis Marchand, Zimmerman, New Jersey assemblywoman Bonnie Watson Coleman, and Mercer County executive Brian M. Hughes. (Photograph by Cliff Moore.)

In 2001, the Association of Members of the Institute for Advanced Study held its biennial conference. Past and current members gathered for a series of social events and public lectures. Speakers for the four public lectures were, from left to right, Islamist scholar Patricia Crone, economist Eric Maskin, theoretical biologist Martin Nowak, and theoretical physicist Nathan Seiberg. (Photograph by Cliff Moore.)

Nine

SYLVAN SETTING
BUILDINGS, GROUNDS, AND WOODS

The Institute for Advanced Study occupies an 800-acre campus in Princeton Township that is situated a mile south of Princeton's main thoroughfare, Nassau Street. Exactly 589 acres of the campus have been preserved from future development. The preserved woods and fields are part of an extensive corridor of green space in central New Jersey and a key ecological link in a network of open space between New York City and Philadelphia.

A walk in the Institute Woods reveals a diversity of species, including aspen, beech, oak, hickory, sweet gum, red maple, and dogwood. It is no wonder that the woods are favored by birders for sightings of migrating songbirds. In spring, the woodland floor is carpeted by spring beauties and yellow celandine along the banks of the Stony Brook. In winter, cross-country skiers explore a delightful landscape of snow and ice.

It seems appropriate that these woods form the backdrop to the Institute for Advanced Study, a place of reflection and discovery. Abraham Flexner once defined the IAS as the place for "curiosity-driven research." "Those who have moved the world have usually followed the will o' the wisp of their own intellectual and spiritual curiosity," said Flexner. Perhaps no one individual better exemplifies this than Albert Einstein. At times, Einstein and others have walked in these woods, alone or with colleagues in deep conversation.

Einstein's remark, "All our science, measured against reality, is primitive and childlike and yet it is the most precious thing we have," and George Kennan's, "True scholars often work in loneliness, compelled to find reward in the awareness that they have made valuable, even beautiful, contributions to the cumulative structure of human knowledge, whether anyone knows it at the time or not," join Flexner's words as inscriptions on the sculpture by Elyn Zimmerman that sits beside the institute pond and, at times, seems part of the natural landscape.

It is no accident that such a setting serves as a backdrop for the life of the mind, and although the campus has expanded since 1939 to include new buildings for the Schools of Mathematics, Natural Sciences, and Social Science; an auditorium; a new library; and a dining hall, every attempt has been made to preserve the sylvan setting with new construction that is on a modest human scale.

The play of natural light in the Fuld Hall common room makes it an easy place to linger. For a part of each day, it is a quiet place in which to sit and read from the selection of international newspapers the IAS provides for its members and faculty.

This mid-1950s aerial view of the Institute for Advanced Study shows Fuld Hall, with its two small flanking wings, and the neocolonial-style office buildings to the rear that were added by J. Robert Oppenheimer in 1948 and in 1953. Trails through the woodlands, located behind the campus, connect to the Princeton Battlefield State Park and the Charles H. Rogers Wildlife Refuge.

Three sets of French doors from the rear of Fuld Hall open onto a patio, with paths leading to the rest of the campus and to the pond and woods beyond. (Photograph by Cliff Moore.)

In 1953, when her husband, Theodore H. Berlin, was a member in the School of Mathematics, Patricia Cleary Berlin created this oil painting of the postwar army-surplus barracks that served as accommodation for members and some faculty since 1946. Those in residence during that time recall the challenges of coal stoves that needed constant stoking during the cold New Jersey winter. The sense of community in the "project" of buildings, such as the one below, was retained when the "village," designed by the Hungarian-born architect Marcel Breuer, was built in 1957. (Courtesy of Dennis Berlin.)

Over the years, member housing has undergone several upgrades, some of which altered Breuer's flat-roofed design. Since this 1985 photograph was taken, there have been further renovations to the institute's member housing complex. (Photograph by Linda Troeller.)

Until the Historical Studies-Social Science Library opened in 1965, the mathematics library in Fuld Hall was the only library on campus. It proved to be a necessary resource during World War II, when restrictions imposed on the movements of "enemy aliens" excluded some IAS scholars from the Princeton University campus. (Photograph by James C. Stephens.)

When Mercer Manor, located to the east of the Princeton battlefield on land owned by the Institute for Advanced Study, was demolished in 1957, the IAS donated part of the building to the State of New Jersey for a battlefield memorial. The columns (at left) that once supported a portico of a Philadelphia residence were designed by Thomas U. Walter, who was also the architect of the dome of the U.S. Capitol Building. When that home in Philadelphia was destroyed around 1900, the portico was brought, partly by boat along the Delaware and Raritan Canal, to Princeton. These photographs show the portico before and after it was disassembled for the move to the battlefield site in 1959. The institute's gift is acknowledged at the site, which was declared a National Historic Monument in 1962.

J. Robert Oppenheimer commissioned architect Wallace K. Harrison to design the modernist-style Historical Studies-Social Science Library and artist Robert R. Wilson to create the cast bronze sculpture, entitled *Nike*, which was unveiled at the building's dedication on April 14, 1965. (Photograph by Kate Ablutz.)

The lower level of the Historical Studies-Social Science Library looks out onto the IAS pond, which provides a peaceful panorama of the changing seasons.

Princeton architect Robert Geddes designed the West Building (above) and the dining hall (below), completed in 1971. Unlike other campus buildings, which look outward to wooded surroundings, these two parallel buildings look inward to a cloister-like courtyard, planted with slender silver birches. The dining hall is the gathering place for the entire IAS community. West Building houses offices for scholars in the School of Social Science and the School of Historical Studies. In recent years, the following new buildings have been named for members of the board of trustees: Wolfensohn Hall, a lecture auditorium and concert venue that was named in 1993; Simonyi Hall, a building for the School of Mathematics that was rededicated in 2000; and Bloomberg Hall, a building for the School of Natural Sciences that was named in 2002. In 2007, Bloomberg Hall was extended to accommodate The Simons Center for Systems Biology, named in recognition of the support of James and Marilyn Simons. (Above, photograph by Adam Ashforth; below, photograph by Bruce M. White.)

Local photographers and artists have often been inspired by the IAS campus. W. Brinton Whitall, a resident of Battle Road, made several studies, including this "Birch Garden in Snow," which captures the stillness of a heavy fall of snow in the courtyard between the dining hall and West Building. (Photograph by W. Brinton Whitall.)

Architects Cesar Pelli and Associates designed the IAS auditorium and School of Mathematics building. The 220-seat auditorium, with its arc roof and pillared entryway, is named for board member James Wolfensohn, and the mathematics building is named for board member Charles Simonyi. The area in front of the buildings includes an outdoor blackboard, a copper-clad fountain, and cherry trees that are a reminder of Caroline Bamberger Fuld. (Photograph by Cliff Moore.)

Sculptor Elyn Zimmerman's tribute to the IAS was installed next to the pond on the occasion of the 75th anniversary celebration in 2005. The words of Abraham Flexner, Albert Einstein, and George Kennan are inscribed on the sculpture's three curved granite benches. (Photograph by Cliff Moore.)

In full summer, the IAS woods envelop walkers in a tunnel of greenery and sparkling light. In 1997, a five-year preservation effort resulted in a permanent conservation easement that protects 589 acres of IAS lands from future development. Popular with local residents for bird watching, hiking, and cross-country skiing, the woods are accessible year-round from the Battlefield State Park public parking area on Mercer Street. (Photograph by Cliff Moore.)

In the early 1950s, Prof. Freeman Dyson and youngsters went to the woods in search of a Christmas tree. The physicist's daughter, Esther Dyson, is the child wearing a hat; her friend Resli Jost sits up front. (Courtesy of George Dyson.)

Nobel laureate P. A. M. (Paul) Dirac (1902–1984), the Lucasian professor of mathematics at the University of Cambridge and a member of the Institute for Advanced Study on numerous occasions from its earliest days, shoulders an ax and sets off for a Veblen-inspired outing to the woods. (Photograph by Ulli Steltzer.)

Princeton philanthropist and IAS trustee Frank E. Taplin Jr. (1915–2003) led the way to preserving 589 acres of institute lands. His personal generosity is acknowledged locally by one of four bronze markers standing in a clearing in the woods, located behind the Thomas Clarke House in Battlefield State Park. Taplin supported several IAS initiatives such as the IAS/Park City Mathematics Institute and the Artist-in-Residence Program. (Courtesy of Public Affairs, Institute for Advanced Study.)

James H. Barbour Jr. and an IAS staff member turned out one Saturday morning in 1980 to register volunteers in a local effort to collect and dispose of gypsy moth larvae on the wooded campus, an attempt to ward off potential damage by the insects. (Photograph by Gary Alvin.)

The trout lily is just one of the wildflowers that appear each spring in the IAS woods. The yellow-flowered plant, whose seeds are dispersed by ants, gets its name from stipples on its leaves that are reminiscent of those on a brook trout. (Photograph by Henry S. Horn.)

Arlen K. Hastings (left) and Imme Dyson have been running together almost every lunch hour since Hastings joined the IAS staff in 1992. Hastings is executive director of the IAS Science Initiative Group (SIG), which promotes science in the developing world. Imme Dyson, whose husband is physicist Freeman Dyson, has run more than 50 marathons and won her age group in most of them. (Photograph by Allen H. Kassof.)

Peter and Helen Goddard are the current residents of Olden Farm, the traditional home of directors of the Institute for Advanced Study. In 2004, mathematical physicist Peter Goddard succeeded Phillip A. Griffiths to become the eighth director. Goddard was previously Master of St. John's College, University of Cambridge. Energetic and gracious hosts, the Goddards play a central role in the social life of both the town of Princeton and the institute. (Photograph by Dinah Kazakoff.)

www.arcadiapublishing.com

Discover books about the town where you grew up, the cities where your friends and families live, the town where your parents met, or even that retirement spot you've been dreaming about. Our Web site provides history lovers with exclusive deals, advanced notification about new titles, e-mail alerts of author events, and much more.

MADE IN THE USA

Arcadia Publishing, the leading local history publisher in the United States, is committed to making history accessible and meaningful through publishing books that celebrate and preserve the heritage of America's people and places. Consistent with our mission to preserve history on a local level, this book was printed in South Carolina on American-made paper and manufactured entirely in the United States.

This book carries the accredited Forest Stewardship Council (FSC) label and is printed on 100 percent FSC-certified paper. Products carrying the FSC label are independently certified to assure consumers that they come from forests that are managed to meet the social, economic, and ecological needs of present and future generations.

FSC
Mixed Sources
Product group from well-managed
forests and other controlled sources

Cert no. SW-COC-001530
www.fsc.org
© 1996 Forest Stewardship Council

Find Your Place in History.